Sport, Spirituality, and Religion

Sport, Spirituality, and Religion: New Intersections

Special Issue Editor

Tracy J. Trothen

MDPI • Basel • Beijing • Wuhan • Barcelona • Belgrade

MDPI

Special Issue Editor
Tracy J. Trothen
The School of Religion and The
School of Rehabilitation Therapy,
Queen's University
Canada

Editorial Office
MDPI
St. Alban-Anlage 66
4052 Basel, Switzerland

This is a reprint of articles from the Special Issue published online in the open access journal *Religions* (ISSN 2077-1444) in 2019 (available at: https://www.mdpi.com/journal/religions/special_issues/religion_sport).

For citation purposes, cite each article independently as indicated on the article page online and as indicated below:

LastName, A.A.; LastName, B.B.; LastName, C.C. Article Title. *Journal Name* **Year**, *Article Number*, Page Range.

ISBN 978-3-03921-830-1 (Pbk)
ISBN 978-3-03921-831-8 (PDF)

Cover image courtesy of Brett Potter.

Contents

About the Special Issue Editor

Tracy J. Trothen is a professor of ethics at Queen's University, jointly appointed to the School of Religion and the School of Rehabilitation Therapy. She is an ordained minister in The United Church of Canada, a certified Supervisor-Educator in Clinical Spiritual Health (CASC), and a Registered Psychotherapist (CRPO). Trothen's areas of research and teaching specialization include embodiment, biomedical and social ethics, Christianity, spiritual health, aging, human enhancement technologies, and sport. Trothen is the author of seven books with the most recent being *Spirituality, Sport, and Doping: More than Just a Game* (2018). Her other books include *Winning the Race? Religion, Hope, and Reshaping the Sport Enhancement Debate* (2015), and the anthology *Religion and Human Enhancement: Death, Values, and Morality* co-edited with Calvin Mercer (2017). She is currently at work on a textbook co-authored with Calvin Mercer tentatively titled *Transhumanism and Religion: An Introduction*. Trothen is a member of the American Academy of Religion's Human Enhancement and Transhumanism Unit Steering Committee.

religions

Editorial

Sport, Spirituality, and Religion: New Intersections

Tracy J. Trothen

School of Religion and School of Rehabilitation Therapy, Queen's University, Kingston, ON K7L 3N6, Canada; trothent@queensu.ca

Received: 12 August 2019; Accepted: 3 September 2019; Published: 23 September 2019

Abstract: Sport, religion, and spirituality intersect in diverse ways. As the body of interdisciplinary scholarly work addressing these intersections increases, more questions and insights are being generated, as evidenced in this collection. Five themes that arise in this Special Issue of *Religions* are identified and explored. Examples from each article are used to develop these themes. To whet the reader's appetite, thought-provoking reflections are offered in this introduction to "Sport, Spirituality, and Religion: New Intersections."

Keywords: sport; religion; spirituality; Christianity; social justice; hope; spiritual emotions

---

1. Introduction

Sport, religion, and spirituality intersect in diverse ways. Scholarship exploring the intersections of sport, spirituality and religion has been growing at a fast pace in recent years (Watson and Parker 2013). This growth may be partly explained by a move away from organized religion towards secularism. Not all countries are seeing a decline of organized religion but such a decline is becoming increasingly evident in parts of the United States and the United Kingdom, the contexts for most of the essays in this Special Issue of *Religions*.

As the body of interdisciplinary scholarly work addressing these intersections increases, more questions and insights are being generated, as evidenced in this collection. The scholars who have contributed articles to this Special Issue include experts in: sports studies; health, human performance and recreation; religion; Christian theology; philosophy; Judaism; rehabilitation therapy; and history. While most of the essays in this issue employ theoretical approaches, Robert Ellis, Terry Shoemaker, Andrew Parker and Mark Oliver also use social scientific research methods. The authors' diverse expertise and research methods yield fascinating new questions and postulations that expand this field of inquiry.

I identify and explore five themes that arise in this Special Issue of *Religions* regarding the intersections of sport, religion, and spirituality: the capacity of sport to achieve ends similar to those ends achieved by organized religions; connections or parallels between sport and Christianity; hope and social justice; the capacity of sport to inspire "spiritual" emotions; and the interpretation or meaning attached to subjective experiences that might seem spiritual or religious in sport. In the hopes of whetting your appetite to read on, I attempt to offer some thought-provoking musings in this introduction to "Sport, Spirituality, and Religion: New Intersections."

1.1. Sports Can Achieve Similar Ends to Religion

Eric Bain-Selbo and Gregory Sapp do an excellent job, in their 2016 book *Understanding Sport as a Religious Phenomenon—An Introduction*, establishing that sports have the capacity to achieve ends similar to those achieved by organized religions (Bain-Selbo and Sapp 2016). Using Ninian Smart's seven dimensions of a religion (Smart 1996), Bain-Selbo and Sapp show that sport can satisfy each of these dimensions. They do not seek to prove that sport actually is a religion but, as Rebecca Alpert puts it in her article concerning baseball's Babe Ruth, "We need not argue whether sports are religions to assert that they can achieve the same ends as systems traditionally defined as religions." While previous publications engage Smart's theory to demonstrate that sport can satisfy the dimensions of a religion (ex. Price 2006; Alpert 2015), four authors in this Special Issue examine new aspects and moments of this intersection adding weight to the claim that sport can achieve the same ends as organized religions.

In her consideration of baseball's Babe Ruth (1895–1948) as a mythic figure (e.g., one of Smart's seven dimensions), Alpert makes the case that Babe Ruth is best understood as a religious icon. Building on the dual strands of Ruth's Roman Catholic religious identity and The United States' response to baseball in the 1920s to the 1950s, Alpert illuminates the function of Babe Ruth as a religious icon in the U.S. American civil religion of baseball. Alpert makes a strong case for baseball being a civil religion, exemplifying the virtues and values of mainstream United States in ways similar to an organized religion. Implications of Ruth's iconic status have been the sacralising of associated objects (e.g., Babe Ruth's team jerseys and baseball equipment) and associated baseball spaces.

Theologian Robert Ellis takes the scope of this Special Issue beyond the United States (as do Andrew Parker and Mark Oliver in their compelling article on sports chaplaincy, and Brett Potter who explores the growing sport phenomenon of parkour) in his article "Sporting space, sacred space: a theology of sporting place." Ellis explores an aspect of Smart's seventh dimension of a religion—materiality—in his consideration of special places in sport. Drawing on United Kingdom examples, Ellis takes us on a tour of special religious and sport spaces including Iona and the Holy Island of Lindisfarne, Durham Cathedral, and the Millennium Stadium. Acknowledging the moral ambiguity of both sport and religion, Ellis posits that the flawed quality of sport and religion does not rule out the "specialness" of some religious spaces and sporting spaces. Ellis concludes "that insofar as sport can be a vehicle for encounter with the divine, it is in no small part because of the part played by its special places in mediating such experiences."

Potter, too, considers the relationship between religion, spirituality, and space in his examination of the global sport of parkour, which he describes as running or moving efficiently though the environments of the globalized city, in "Tracing the Landscape: Re-enchantment, Play, and Spirituality in Parkour." Potter sees parkour and its offshoot, free-running, as "like" a religion insofar as parkour exhibits aspects of religion such as pilgrimage, and the promotion of virtues and pillars that are similar to some of those in Buddhism and Hinduism.

Bain-Selbo suggests that sport can function similarly to religion by providing the possibility of so-called spiritual or religious emotions. In his article, "Affect Theory, Religion, and Sport," Bain-Selbo builds the argument that affect generated in sport may be the same as affect generated in religion. Using Donovan O. Schaefer's understanding of affects as "the flow of forces through bodies outside of, prior to, or underneath language," Bain-Selbo reflects on the possible meanings of affect in sport and the need for critical interpretation of these affect experiences. He builds a strong case for affects as ends in religions and similarly in sport. The fact that many people who experience powerful affect in sport do not name these experiences as spiritual or religious is likely at least partly due to contextual labelling. Even though the same words may not be used to describe strong emotions experienced in religion and in sport, Bain-Selbo suggests that religion and sport can achieve similar or even the same strong affects.

In his essay "Deconversion, Sport, and Rehabilitative Hope," Terry Shoemaker presents findings from his qualitative and ethnographic research on post-evangelical followers in the northern United States portion of the "Bible Belt." No questions specific to sport were asked in the 65 interviews but 15% of the respondents referred to sport in reflecting on their experiences of leaving a conservative evangelical church. Shoemaker's research suggests that sport may have the capacity to fulfil the ends of relationality and hope sometimes more effectively than religion. He concludes that sport intersects—often in constructive ways—with deconversion from conservative evangelical Christian churches for some people.

These essays together make a strong case for sport offering many of the same ends that may be sought in organized religions. The fact that these authors chose their topics for this Special Issue independently may suggest that this theme is attracting growing interest and conviction.

1.2. Connections or Parallels between Sport and Christianity

Alpert, Ellis, Shoemaker, and Andrew Meyer all draw connections or parallels between sport and Christianity. While this theme is not new in scholarship regarding religion and sport, the points of intersection identified are new or the authors offer new insights.

Alpert's exploration of Babe Ruth emphasizes Ruth's role in bringing Roman Catholicism to mainstream America. Alpert helps the reader to understand how Babe Ruth exemplified aspects of the Roman Catholic religious commitment to social justice, particularly for children and the poor. She surmises that Ruth's huge successes on the baseball diamond gave more power and perhaps mystique to his Roman Catholic identity. Alpert provides a number or rich examples that illustrate this connection, personified in Babe Ruth, between baseball and Roman Catholicism at a time when the reputation of baseball was precarious.

Meyer uses the biblical source to shed light on another high-profile athlete, cyclist Lance Armstrong, and his dramatic fall from grace. Meyer reflects on the possible meaning of Armstrong's doping scandal. Using the Isaiah narratives, Meyer proposes that this biblical source can reveal redemptive meaning in Armstrong's narrative of prelapsarian glory followed by a fall of almost biblical proportions. Through being forced into exile, Meyer surmises that Lance may have gained the possibility of redemption not as an elite cyclist but, more importantly, as a person. This exploration of a parallel between the Isaiah narratives and Lance Armstrong's narrative helps the reader to see both Armstrong and this biblical story from another perspective.

Shoemaker looks at another way in which sport intersects with Christianity, inviting us to consider the relationship of sport to the deconversion experiences of some who have left conservative Christian evangelical churches. He predicts that further research may support the role of sport in providing alternative identities, relationships, and rehabilitative hope for many who decide that their evangelical Christian community no longer fits with their identities, convictions and desires.

Jeffrey Scholes examines the relationship between racialization and reactions to displays of Christian faith on the gridiron, in his ground-breaking article "Pray the White Way: Religion Expression in the NFL in Black and White." Scholes makes the reader pause and ask critical questions regarding how we evaluate some theological interpretations as superior to others. In this re-consideration he finds that the types of theologies that are judged more harshly are African American theologies. Scholes builds a persuasive case. This article will change how you see athletes, and especially black National Football League (NFL) players, who attribute a win to God.

Through engagement with Christian theological concepts including incarnation, sacrament, and Trinity, Ellis suggests that there are important similarities between the qualities and experiences of Christian and other religious special places and sporting special places. In so doing, he adds weight to arguments against binary approaches to the sacred and the profane. From a Christian theological perspective, Ellis proposes that God may mediate "saving presence and grace" in any place, including sports' spaces.

These papers develop new angles and even planes in the intersection of Christianity and sport. The authors give us new lenses through which to see sports and athletes in relation to Christianity.

1.3. Hope and Social Justice

All the authors in this Special Issue of *Religions* address hope and most connect hope to social justice. The authors see both sport and religion as having the ambiguous potential both to challenge and to reinforce socially normative patterns of power, privilege, and marginalization.

Sport researchers Parker and Oliver explore new ground in "Safeguarding, Chaplaincy and English Professional Football." In a qualitative study, Parker and Oliver investigated the role of sports chaplains in safeguarding the welfare of vulnerable elite youth footballers in England. This safeguarding can be understood as offering hope for the well-being of vulnerable players, and thus promoting social justice in the English football context. Their research suggests that the perception of the chaplain as removed from the team and safe, may give the chaplains protective potential; sports chaplains seem to be particularly trusted figures and so have a greater opportunity to protect these young footballers from abuse. As Parker and Oliver point out, chaplains have a unique role in elite youth football in that they are able to prioritize the well-being of the players over their sports performance. The authors implicitly suggest that religiously informed values and behaviors can be more person-oriented and protective than elite sports values.

Scholes addresses social justice in the NFL in his critical analysis of elite sport and of normative white Christian discourse in the United States. Scholes' important consideration of racialized evaluations of religious expression in professional American football gives one pause. Is the denigration of providential and hope-filled theological declarations in sport (e.g., God decided that our team should win tonight) a social and theological justice issue? I am convinced now that it is. The question of whether God cares about football connects in no small way to big political and economic questions. When Scholes quotes journalist David French as saying "to get God out of football, the anti-religious crowd would need to get the football players out of football," as a Canadian I immediately thought of Quebec's recent secularism law, Bill 21, which forbids religious symbols in the public sector (passed into law on 16 June 2019). Legislated or rule-based attempts to address religious and secular diversity by denying such expression arguably denies the diversity of human faith commitments and world views. At the same time, philosopher Randolph Feezell's point that faith expressions ought to be held fallibly in a pluralistic context, must be held in tension with public faith expressions (Feezell 2013). Scholes' article is a strong reminder of the insidiousness of racism and the importance of faith fallibility. It can be easy to judge players who seem to declare their faith as infallible but what about the critics who assume that their judgements regarding players' faith expressions are infallible?

Hope is a significant theme in Meyer's article. Meyer sees Lance Armstrong as a complicated sports icon, fallen hero, and potential symbol of the promised land. Meyer is interested in the ambiguity represented by Armstrong in his "Livestrong" charitable work and his advocacy for those battling cancer as he himself did. For Meyer, Armstrong represents human fallibility and the redemptive possibility that is promised by religions such as Christianity. Meyers concludes, "there is hope in despair" and, more profoundly and even controversially, it may be that true hope requires despair. For those who ask where the social justice is in Lance Armstrong's gradual emergence from exile, you will find Meyer's article thought provoking.

Shoemaker's article is more immediately about hope than social justice. However, if an aspect of social justice is the re-claiming of one's faith and sport identities and values, then this article is also about social justice for those who experience marginalization because of what they believe. In his examination of post-conservative Protestants including evangelicals, fundamentalists, and Pentecostal followers, Shoemaker argues that "sport can offer a rehabilitative hope for familial fracturing caused by religious deconversion, a spatial opportunity in a restrictive milieu to expand one's social network, and a space to align with contemporary social issues." Shoemaker's research shows that sporting interactions can be disruptive to faith identities by interacting with others with different racial, sexual, and faith identities. In other words, sport experiences can contribute to a choice to deconvert. But sport can also assist in the healing of relational, including familial, breaks that are caused by "religious shifts away from evangelicalism." In the study, sport was also identified as a place in which people who experienced deconversion could connect with others who share similar emergent values. Sport may provide rehabilitative hope in the work of deconversion and self-discovery.

In linking American civil religion and Babe Ruth's Catholic faith identity, Alpert shows that Ruth was an icon of hope for succeeding against the social odds, especially in the context of the two world wars and the Depression of the 1930s. People from all social classes could identify with him in some way. His Catholic faith was potentially marginalizing but he embraced it. Alpert emphasizes that it was not his confessions of bad behavior and apologies as much as his sometimes-overlooked championing of those on the margins, including children and the poor, that may have been the clearest expression of his faith. Alpert shows how Babe Ruth, religious icon and saviour, helped to illuminate the social justice aspect of Roman Catholicism and the persistent possibility of hope for the downtrodden.

Ellis also addresses hope as it is associated with special places in religions and sport. He describes an almost palpable sense of divine love and awe that is generated in some sport places that can help to bring us a strong sense of spiritual presence. This sense of awe and "wow" is not mitigated by human moral fallibility. Rather, as Ellis persuasively argues, the hope that these special places can inspire may be more sacramental in nature, an unfathomable generous unveiling of the divine. Ellis invites us to be critical but open to holy experiences in special places that are not necessarily religious places. His article makes me wonder how the emotional and spiritual "wow" factor may be related to good (or bad) works.

Potter shows us that parkour can be a global way to raise the valuing of the embodied person and the environment, and specifically the urban environment as it connects to human spirituality. He suggests the concept of religious poesis for understanding the transformation of the body through engagement with the urban landscape. Potter concludes that parkour and free-running provide us with a way forward towards ecojustice: "Perhaps parkour, with its emphasis on embodied experience of the landscape, can help transform this [urban] desert into an oasis. An 'enchanted' view of not only nature, but the urban landscape itself as full of potential for beauty and grace, could perhaps be the basis for a renewed ecological consciousness." As Potter points out, social justice must include the creative rehabilitation of the earth.

Hope is often explicitly linked to social justice in the above essays. The authors agree that hope is not restricted to religion but is found by many, and sometimes more easily, in sport than in religion. As these authors demonstrate, sport and religion are ambiguous. They share the capacity for generating hope and modeling social justice, and they share the capacity for abuse and oppression.

1.4. Sport and (Spiritual?) Strong Emotions

This theme is perhaps the least obvious and most limited in terms of how much is said directly concerning emotions in the articles. Yet, I think emotions are an important theme since they are mentioned often, even if briefly or implicitly, in the articles by Alpert, Ellis, Potter, Scholes, Meyer, Shoemaker, and Bain-Selbo.

Alpert demonstrates that many of Babe Ruth's followers made him into a religious icon and, in the same way as a religious icon, Ruth redeemed baseball's "awe-inspiring power and beauty" from revelations of gambling and violence. Ellis writes about the "wow" moment of encountering special sport and religious places. In his exploration of parkour and free-running, Potter also makes implicit reference to the "wow" factor of the "enchanted" urban landscape "as full of potential for beauty and grace." Scholes looks at religious displays on the gridiron; no one would argue that these displays are not partly fueled by strong emotions of ecstasy, awe, hope, and gratitude. Meyer's essay on Lance Armstrong is laced with strong emotions including despair, suffering, and hope. Shoemaker distills three types of rehabilitative hope in sport for people who are experiencing the emotional stress of deconversion. What is the meaning of these strong emotions? Why are strong emotions so consistently associated with the intersection of sport, spirituality, and religion?

Bain-Selbo helps us to probe these questions about emotions, religion, and sport by engaging affect theory. His contention that there are no irreducibly religious emotions caused me to ponder further the distinction between religious and spiritual in regard to affects. I agree that "the physiological changes (the felt experience) of the religious experience are in fact religious to the extent that they are interpreted religiously—they are not intrinsically religious." And I would take Bain-Selbo's argument a step further: I wonder if some of these emotional responses might be considered "intrinsically spiritual." By intrinsic, I mean that these emotional responses might possess an irreducible spiritual quality that is not restricted to any particular context, including religions (Pargament 2013b).

I have written about sport as a place where some people discover the sacred, building on research by psychologist Kenneth I. Pargament and his colleagues (Trothen 2018). I cannot do their work justice in this very brief reflection but I will summarize a few key points and leave you, the reader, to investigate their findings more closely should you choose. Pargament and his colleagues have found that people can discover the sacred in any number of life experiences and/or objects. They also have found that these experiences of the sacred cannot be fully explained by or reduced to psychological needs, or desires, or any other motivation. In his research, Pargament observed six implications for one's everyday life stemming from the discovery of the sacred (Pargament 2013a; Pargament et al. 2017) and that the experience of discovering the sacred has at least three qualities: transcendence, ultimacy, and boundlessness (Pargament 2007). One of the implications of discovering the sacred is the experience of strong emotions that may be considered spiritual emotions (Pargament 2013a) when they occur conjointly with the other five implications and the three or more spiritual qualities. It may well be that some people discover the sacred in sport and experience emotions that can be understood as (irreducibly?) spiritual.

I suspect that these questions about strong emotions, spirituality, religion, and sport will attract increasing attention. The emotional intensity in sport can be expressed in life-enhancing as well as destructive ways, as we have seen in large fan crowds. If these emotions have a spiritual dimension for at least some people, we need to better understand the relevance and meaning of this spiritual dimension if we want to assist people to express these emotions constructively (Trothen 2018).

1.5. The Interpretation of Meaning: Is Everything Spiritual or Religious Because I Say So?

The understanding of strong emotions in sport is part of a bigger question: how do we interpret the meaning and value of sport and religion, and/or spirituality, when they intersect? The articles by Bain-Selbo, Ellis, Potter, Scholes, and Parker and Oliver, suggest this question.

Bain-Selbo repeats Wayne Proudfoot's caution that "we must be careful not to protect claims of affect and religion from rational inquiry." We should not accept that any strong emotion experienced in sport (or in any context) is a religious affect experience just on someone's say-so. We need more critical interpretation. Bain-Selbo pushes this caution further and asks if there are no sui generis religious affect experiences, then is it not possible to have emotional experiences in sport that are similar to or the same as emotional experiences in religion? If so, then it may be that those who have bought into a false dichotomy between the religious and the secular have misapprehended authentic religious or spiritual experiences in sport. As he puts is, "If the participants in the sporting event had a different understanding of what religion is or what a spiritual experience is, perhaps they more likely would apply these terms to describe their experiences and, thus, those experiences legitimately could be considered religious or spiritual." In raising the possibility that some people have what have been understood as religious or spiritual experiences in sport, Bain-Selbo points to a looming question: how well do we understand or even recognize religion or spirituality?

Ellis and Potter also urge us to question a sharp distinction between the so-called sacred and the so-called profane, in their respective explorations of special places in sport and religion, and the parkour participant in relation to the urban landscape. Cautioning against an assumption that any and all places *are* sacred or spiritual, Ellis nonetheless argues that a place *can be* sacred or spiritual. He rejects a pantheistic understanding of these special places, arguing instead that while God can mediate love and other divine actions through special places, we need to critically explore the meaning of our experiences in these places. Ellis proposes theological concepts that can assist in this critical evaluation and discernment of spirituality in particular places. Ellis concludes that "[w]e should resist any tendency to separate 'sacred space' too rigorously and find a balance between the rejection of the notion of sacred space and the banal affirmation that every space is always sacred."

Potter echoes Ellis by challenging the idea that only particular spaces are associated with the spiritual: "This tendency to separate out the sacred and the secular is precisely what is called into question by emergent practices such as parkour and free-running." Potter sees parkour as reconfiguring (urban) space by not confining "enchantment" to "'sacred' spaces like the stadium, gallery, or church." So, is enchantment the same as spirituality? How should we understand the ability of parkour to reconfigure space and defy any distinction between the sacred and the secular? How is this proposal different from Ellis'? Ellis puts an emphasis on divine revelation and human response whereas Potter may see human agency, expressed in parkour or free-running, as mainly or solely responsible for the reconfiguration of urban space into sacred space. The question of how we interpret and discern sacredness comes to the fore in these two papers. I encourage you to read both and see what you think.

Scholes, too, raises the question of interpretation in looking at providential faith convictions expressed by African American NFL players. Is God's hand in the winning touchdown? Is a completed pass providential, a manifestation of the sacred? Who gets to decide which faith interpretations are acceptable? How do we promote fallible religious claims in the public sphere without also promoting racist evaluations of players' religious beliefs? Scholes reminds us that our own judgements of religious displays can manifest promote a sort of infallibility and arrogance.

The question of what makes a sport experience spiritual or religious also emerges implicitly in Parker and Oliver's article. Young footballers may perceive chaplains as invested not in their sport performance primarily but in their well-being. The chaplains seem not to be interpreted as being owned by the team but as being accountable and committed to their faith. Thus, (I would add) rightly or wrongly, the chaplains may be seen as safer and more trustworthy. Again, a key question concerns how we interpret the meaning of religious or spiritual and where we discern religious or spiritual qualities.

2. Concluding Remarks and Invitation to This Special Issue

I have made a mere beginning in distilling some of the significant themes raised in this Special Issue of *Religions* on "Sport, Spirituality, and Religion: New Intersections" and I urge the reader to look for additional themes. These essays help us to understand the complexities of the intersection of sport, spirituality, and religion. While these essays do not explicitly venture outside of the United States and the United Kingdom, with the notable exception of Potter's intriguing article on parkour, I suspect that many of these themes and questions could be extended into additional cultural contexts. I urge you to consider how these articles might generate further insights and direct engagement with more diverse contexts. I am sure that you will find yourself stimulated and challenged by this impressive collection. Enjoy!

Funding: This research received no external funding.

Conflicts of Interest: The author declares no conflict of interest.

References

Alpert, Rebecca T. 2015. *Religion and Sports: An Introduction and Case Studies*. New York: Columbia University Press.

Bain-Selbo, Eric, and Gregory D. Sapp. 2016. *Understanding Sport as a Religious Phenomenon—An Introduction*. London: Bloomsbury Academic.

Feezell, Randolph. 2013. Sport, Religious Belief, and Religious Diversity. *Journal of the Philosophy of Sport* 40: 135–62. [CrossRef]

Pargament, Kenneth I. 2007. *Spiritually Integrated Psychotherapy: Understanding and Addressing the Sacred*. New York: Guilford.

Pargament, Kenneth I. 2013a. Searching for the sacred: Toward a non-reductionist non-reductionist theory of spirituality. In *APA Handbooks in Psychology, Religion, and Spirituality: Context, Theory, and Research*. Edited by Kenneth I. Pargament, Julie J. Exline and James W. Jones. Washington: American Psychological Association, vol. 1, pp. 257–74.

Pargament, Kenneth I. 2013b. Spirituality as an irreducible human motivation and process. *The International Journal for the Psychology of Religion* 23: 271–81. [CrossRef]

Pargament, Kenneth I., Doug Oman, Julie Pomerleau, and Annette Mahoney. 2017. Some contributions of a psychological approach to the study of the sacred. *Religion* 47: 718–44. [CrossRef]

Price, Joseph L. 2006. *Rounding the Bases: Baseball and Religion in America*. Macon: Mercer University Press.

Smart, Ninian. 1996. *Dimensions of the Sacred: An Anatomy of the World's Beliefs*. Berkeley: University of California Press.

Trothen, Tracy J. 2018. *Spirituality, Sport, and Doping: More Than Just a Game*. Springer Briefs Sport and Religion Series; Basel: Springer International Publishing.

Watson, Nick J., and Andrew Parker. 2013. Sports and Christianity: Mapping the Field. In *Sports and Christianity: Historical and Contemporary Perspectives*. Edited by Nick J. Watson and Andrew Parker. New York: Routledge, pp. 9–88.

religions — MDPI

Article

Safeguarding, Chaplaincy and English Professional Football

Mark Oliver and Andrew Parker *

School of Sport and Exercise, University of Gloucestershire, Gloucester GL2 9HW, UK;
mark@friendlydevelopment.co.uk
* Correspondence: andrewparkerconsultingltd@gmail.com

Received: 3 July 2019; Accepted: 11 September 2019; Published: 22 September 2019

Abstract: In recent years, English professional football has been rocked by allegations of historical sexual abuse and safeguarding concerns around young players. This paper examines the potential contribution that sports chaplains can make to the specific welfare needs of elite youth footballers within the wider context of safeguarding practices and protocols. Comprising a small-scale, sociological study involving welfare personnel at English Premier League and English Football League Championship clubs, the paper identifies the scope and potential of sports chaplaincy in relation to the practical outworking of safeguarding policy. Findings reveal that elite youth footballers face a number of pressures specific to the highly competitive environment in which they work and that various safeguarding issues routinely arise amidst these pressures. The paper concludes by suggesting that sports chaplains are ideally placed to provide safeguarding and wider welfare support to young players as a consequence of their independence from team management structures and their prioritization of holistic care above performance-related issues.

Keywords: safeguarding; elite youth sport; English professional football; qualitative research

1. Introduction

Association (professional) football in the UK has been subject to an unprecedented number of allegations of abuse in recent years, the majority of which date back over several decades. These historical cases have impacted a significant number of individuals and organisations from grassroots to elite level with the integrity of coaches and officials coming into question around safeguarding issues (see, Conn 2018; Morris 2019; Taylor 2018; Wallace 2019). As a consequence, governing bodies of sport in the UK have been stimulated to implement robust safeguarding policies and procedures in order to provide safer, nurturing environments in which young people can thrive (see, for example, The Football Association 2015, 2017).[1]

Sports chaplaincy has developed markedly during the same period and is represented in a large number of professional football clubs in the UK providing holistic pastoral and spiritual care both for athletes and wider employee groups. A small body of research has emerged detailing the working lives of chaplains within this context yet little has been written about how and to what extent sports chaplains might specifically contribute to the safeguarding and wider welfare support of elite youth footballers. This paper aims to address this shortfall by exploring some of the pressures and associated safeguarding issues that exist within this particular context and by investigating the extent to which chaplains may have the opportunity to make a meaningful contribution in this area.

---

[1] For further discussion on the role of chaplaincy provision within the broader context of athlete welfare and wellbeing see Hemmings et al. (2019a).

The paper begins by discussing recent welfare reforms in UK professional sport especially those surrounding safeguarding. Following this, we explore the roles and responsibilities of chaplains in professional football and some of the specific pressures faced by elite youth players. Drawing upon the findings of a small-scale, sociological study into safeguarding provision at six English professional football clubs, the paper reveals that elite youth footballers often face a range of performance-related pressures from coaches, parents and peers which, if not dealt with appropriately, may lead to wider welfare concerns. Findings also reveal that safeguarding co-ordinators were unanimous in their recommendation of club chaplains and their unique position as trusted figures who can provide pastoral support independent of those in decision-making positions. The paper concludes by suggesting that sports chaplains can contribute meaningfully to both the safeguarding and wider welfare needs of elite youth footballers.

2. Safeguarding and UK Sport

In the UK at least, the prominence of safeguarding has increased markedly both in sport and wider social contexts during the 21st century.[2] Awareness of abuse has been raised via media coverage of allegations involving high-profile individuals and institutions, whilst child protection training, policies and praxis such as criminal record checks have also placed safeguarding in the public consciousness.[3]

In her review of duty of care and safeguarding procedures in sport, ex-Paralympian (Baroness) Grey-Thompson (2017) argues that success in sporting contexts is often dependent upon prioritizing participant safety, wellbeing and welfare. This reflects Sport England (2016) current strategic aim to see wider, holistic personal development outcomes prioritised over and above more traditional objectives such as increased sporting participation and elite success. However, this has not always been the case and prior to the 1980s sport was often viewed as a somewhat disparate and isolated entity, lacking accountability to wider social structures (Brackenridge and Rhind 2014). The Child Protection in Sport Strategic Plan developed between the UK National Society for the Protection of Cruelty to Children (NSPCC) and Sport England, states that the focus for safeguarding in sport should be on elite athletes as they are amongst the most likely to face discrimination, exclusion and abuse due to the unique pressures associated with such levels of performance (Sport Safeguarding Partnership 2016).

In terms of the propensity of sporting environments to breach safeguarding norms, the culture within English professional football is a case in point. In recent years, the reputation of football has been significantly impacted by allegations of historical sexual abuse attributed to an institutional culture of power and control.[4] As a consequence, football's governing bodies have sought to develop and implement robust safeguarding policies and procedures in a co-ordinated attempt to prevent the recurrence of such events, a strategy which has included significant investment to embed specialist safeguarding provision across all levels of the sport and to promote a collaborative approach to wider welfare concern in order to cater for the holistic development of players. This work has stimulated improved safeguarding policies and training and has included the launch of the Football Association's (FA) Child Protection strategy in 2000 (Brackenridge et al. 2004, 2005).

Concurrently, the FA became the first UK national governing body of sport to commission a major investigation into child protection which was tasked with evaluating the impact of these policy measures on the culture of the professional game (see Pitchford et al. 2004). Featuring over 300 amateur football clubs, one of the key findings of this work revealed patterns of failure around the safeguarding of youth participants (Pitchford et al. 2004; Brackenridge et al. 2005) which subsequent policies have

---

[2] In the UK, the term 'safeguarding' refers to the protection of the health, wellbeing and human rights of children/young people and adults at risk, enabling them to live safely and free from abuse and/or neglect. For more on the importance of safeguarding within the context of broader sport for development agendas see Giulianotti et al. (2019).

[3] This section of the paper draws upon some of our previously published work (see Hemmings et al. 2019b).

[4] An Independent Review into Child Sex Abuse Allegations in Football is currently being led by Clive Sheldon QC. The Review into what the Football Association and Clubs knew and did about allegations of child sexual abuse between 1970 and 2005 has run alongside ongoing criminal investigations (see BBC 2018; Roan 2019).

attempted to address (see Mountjoy et al. 2015; Sport Safeguarding Partnership 2016; Grey-Thompson 2017; Ong et al. 2018).

Other safeguarding issues reported in the study included the aggressive touchline behaviours of parents and allegations of underage alcohol consumption. Although this research focused on grassroots football, resulting data suggested that abusive behaviours increased proportionately with ability level (Brackenridge et al. 2004; Brackenridge and Rhind 2014), a pattern validated by more recent research findings (International Centre of Ethics in Sport 2016; Grey-Thompson 2017; Ong et al. 2018).

Of course, it is well-established that elite youth athletes are more likely to experience overtraining, performance-related stress and excessive mental and emotional pressure (Mountjoy et al. 2015; Manley et al. 2016; Parker 2016; Parker and Manley 2016, 2017). Likewise, it is acknowledged that a variety of safeguarding issues can arise in elite youth sport as a consequence of the dependency dynamics in play and the intense nature of coach-athlete relations (Brackenridge et al. 2005; International Centre of Ethics in Sport 2016; Grey-Thompson 2017). These issues can be perpetuated by the prioritization of performance over pastoral care (Ong et al. 2018), an area in which some coaches appear to lack suitable skills and knowledge (Grey-Thompson 2017). According to Mountjoy et al. (2015), recent studies reveal that the coach is not always the perpetrator of abuse, particularly in the case of bullying and 'hazing', where peers are most likely to act as instigators (see Kavanagh 2014, Brackenridge and Rhind 2014). Informal 'banter' between participants is commonplace in elite sport (Grey-Thompson 2017) and may often mask psychological and emotional vulnerabilities (Ong et al. 2018), with players possessing a wide range of coping strategies. The harsh realities of such interaction have long been recognised within the context of youth football (Pitchford et al. 2004; Brackenridge et al. 2004); examples including the verbal chastisement and the physical 'punishment' of players by coaches (see Parker 1995). For this reason, Grey-Thompson (2017) urges coaches to identify instances where 'banter' leads to bullying and to eradicate the damaging elements of such behaviours. In turn, she acknowledges the imbalance of power between coach and athlete, arguing that these dynamics can often be viewed as 'the norm' in elite sport; a rite of passage that participants are expected to endure given their lack of influence in terms of culture change (see Parker 2001, 2006).

Research and policy around safeguarding have led to a number of practical recommendations for creating safer sporting environments. It is suggested, for example, that football (and sport more broadly) should change its overall approach by adopting a child-centred player development model suitable for the unique culture in which young people are based (Ong et al. 2018; Pitchford et al. 2004; International Centre of Ethics in Sport 2016). Moreover, elite players (of all ages) should be supported during transitions into and out of sport, with player welfare being embedded in robust induction and de-selection processes (Grey-Thompson 2017). There is evidence in English professional football of youth academy staff being encouraged to place greater emphasis on the establishment of trust and rapport with players in order to better understand their behaviours and facilitate meaningful conversations (Ong et al. 2018). For example, recent developments at Manchester City Football Club have witnessed collaborations between medical, safeguarding, psychology and sports science teams in order to enhance player welfare (Ong et al. 2018). In turn, Grey-Thompson (2017) highlights the importance of referring athletes to 'independent and confidential support services' as players often fear jeopardising their careers if they raise welfare concerns with coaching staff. Such holistic practices clearly present opportunities for helping professionals—including sports chaplains—to bring together their individual and collective skills and expertise in relation to wider welfare support (see Hemmings et al. 2019b).[5]

---

[5] For more critical discussion of the implications of the increased prevalence of safeguarding protocols on elite sport see Baker et al. (2017) and Barker-Ruchti (2019).

3. Chaplaincy and UK Sport

Sports chaplaincy in the UK mirrors many of the roles and responsibilities of generic chaplaincy, i.e., pastoral/spiritual care, crisis support and personal discipleship (Parker et al. 2016). These duties must be underpinned by the chaplain's ability to apply them to the specific context in which they find themselves (Slater 2015; Swift et al. 2015; Caperon et al. 2018). It would be fair to say that English professional football has featured large in the development of UK sports chaplaincy acting as a catalyst for the emergence of such provision across elite sport and serving as a training ground for numerous practitioners (see Heskins and Baker 2006). As is the case for many chaplains operating in elite football, those featured in the present study all occupied positions of marginality within their host institutions (see, for example, Roe and Parker 2016), a culturally hostile environment renowned for its occupational instability and hyper-masculine workplace practices (Roderick 2006; Parker 2001, 2006). Whilst at first glance marginality might be viewed as something of a disadvantage, Pattison (2015) has argued that as an accepted part of chaplaincy practice, such role ambiguity may in fact bring with it a whole series of benefits in relation to organisational access and influence.

Chaplaincy provision has increased significantly in English professional football in more recent years. For example, the majority of English Premier League clubs now have a sports chaplain working as part of their broader holistic support networks. In this role, chaplains offer practical, spiritual and pastoral support, as well as providing assistance with wider lifestyle issues (Gamble et al. 2013; Nesti and Sulley 2015). At times, players and staff may be reticent to engage with such support on account of the mistaken perception that the chaplain is there to proselytise their faith and research suggests that club managers may be influential in this respect (Roe and Parker 2016). In their study of the relationship between sports chaplains and sport psychologists in the English Premier League, Gamble et al. (2013) found that chaplains felt that their role may be either empowered or disempowered depending on the attitude of a team manager towards them. Nesti and Sulley (2015) emphasise the significance of these facets of organisational culture, arguing that they often provide an unspoken set of rules that represent the club's collective morality, allowing or prohibiting certain types of behavior. However, Nesti and Sulley (2015) go on to argue that the chaplain's ability to function as an effective, yet distinct, part of a club's support network is fundamentally reflective of the depth of relationships that they build with players and staff.

In elite sports settings, the work of the chaplain often enhances broader welfare provision identifying such opportunities through cross-departmental dialogue (Hemmings and Chawner 2019) and by responding flexibly to the needs of those around them. It is argued that such practices can contribute to improved athlete performance but it is the distinct spiritual nature of the chaplain's work which enables them to place value on each individual, regardless of their perceived worth to club officials, coaches or to the team itself (Gamble et al. 2013). Given the degree of uncertainty surrounding the personal and professional lives of elite youth footballers and the social and psychological challenges that they face (Professional Footballers' Association 2017), Baker (2006) argues that sports chaplains should be mindful of the specific needs of young players. One of the ways in which this often plays out within the context of English professional football is via the neutrality and independence of the chaplain's position in relation to club hierarchies and decision making processes (Weir 2016). Unlike other support services, chaplains are not typically employed by the clubs that they serve and do not therefore report directly to management on matters of welfare or performance (Roe and Parker 2016). Professional footballers can be reluctant to speak to those with decision-making responsibilities (i.e., club management and coaching staff) about personal issues due to the impact that this may have on team selection and/or career progression (Nesti and Sulley 2015). A central strength of chaplaincy is the prioritization of player personal need over and above that of team performance (Comfort 2006; Baker 2006). Although arguments prevail around the negative implications of this independent if marginalized (and somewhat isolated) position (Knight 2006; Comfort 2006), it is the central contention of this paper that such a role is fundamental to the chaplain's capacity to contribute to the safeguarding of elite youth players.

4. Methodology and Method

Professional football academies facilitate a comprehensive programme of competition, training, education and support for boys and young men in structured age-groups between nine and 23 years of age (The English Football League 2018). Depending upon the nature of the academy in question (four category levels of academy exist based on the standard of resources and facilities on offer, i.e., staff/coaching, sports science and education), players progress through part-time, full-time or hybrid models of training and education negotiated between club, parents and school (The Premier League 2018). At 16, those young players who are to be retained by their host club are offered a formal 'scholarship' which comprises a full-time footballing contract accompanied by part-time academic study, typically amounting to the completion of an advanced apprenticeship in sporting excellence, alongside their holistic development as athletes. This includes time allocated to focus on life skills including workshops delivered by club staff and external agencies who specialise in a variety of topics relating to the pressures often faced by young players such as gambling, substance misuse, social media and personal relationships (Professional Footballers' Association 2017).

The research reported here was carried out between November 2017 and March 2018 and the primary method of data collection was semi-structured interviews. The aim of the research was to explore whether and to what extent sports chaplains could make a meaningful contribution towards safeguarding in professional football. Interviews took place with those responsible for wider welfare support in football clubs which hosted a sports chaplain[6] and were carried out within six clubs; three from the English Premier League and three from the English Football League Championship, all of which were located in urban conurbations across the north and south of the UK. The Premier League and Championship clubs were represented by their Heads of Safeguarding and Heads of Education and Welfare respectively, in accordance with each league's safeguarding policies. In order to ensure that an appropriate data set was obtained, purposive sampling was used in relation to participant selection (Bryman 2015).

Prior to the onset of the research, ethical approval was granted by the University of Gloucestershire. All respondents were contacted directly via email in the first instance and each was issued with an information sheet and consent form regarding the collation of data prior to interviews taking place. Interviews lasted between 30 and 40 min and were audio recorded and transcribed in full. Four were held in-person at the host clubs of respondents and two were conducted via Skype. Interview questions were derived from the background academic and policy literature in the area and framed in line with the specific focus of the research itself. Respondents were asked to reflect on their roles and responsibilities relating to wider welfare support including how they helped players deal with the specific pressures that may arise in elite youth football. They were also invited to discuss the types of safeguarding issues that may emerge within this context. Respondents were then asked to share their understanding of the sports chaplains' role in their respective clubs and to comment on whether they thought there was potential for chaplains to contribute (or contribute further) towards some of the pertinent safeguarding and wider welfare issues in play. Finally, respondents were invited to share any further relevant information. The questioning style during interviews was open-ended and, where necessary, further probing took place to clarify responses. Discussions focused on a range of issues surrounding the holistic wellbeing and/or performance of elite youth players. In line with conventional practice concerning processes of respondent validation, all participants were offered the opportunity to review their interview transcripts for accuracy (Bryman 2015). In the interests of anonymity, pseudonyms have been used throughout the present discussion.

A grounded theory approach to data analysis was deployed in line with Strauss and Corbin (1998). Grounded theory allows for the systematic analysis of data through a process of open, axial and selective coding and the formation of a conceptual framework that facilitates the presentation of

---

6   Of the six chaplains whose clubs featured, two were ordained ministers and the remainder were lay chaplains.

participant experiences from their own perspective (see Charmaz 2000, 2014). To this end, data were analysed in four stages. Firstly, transcripts were read in full to gain a comprehensive overview of the data. Secondly, each transcript was individually coded and indexed whereby a capturing of the different aspects of participant experience took place. Thirdly, these experiences were then categorised into a number of over-arching topics. The final stage of analysis involved the formal organisation of these topics into generic themes two of which provide the focus for the remainder of our discussion. The first highlights the kind of safeguarding issues that respondents found themselves dealing with while the second explores how and to what extent chaplains may contribute to safeguarding and wider welfare support within youth football. Prior to our detailed consideration of these themes, we present a snap-shot of the occupational profiles of each of our respondents and their histories in the professional game.

Edith Ford—Head of Safeguarding, Sunnyside Rovers (English Premier League).

Edith Ford has worked at Sunnyside Rovers for more than 20 years, fulfilling various roles including Women and Girls' Football Development Officer. In 2004 she became the club's Head of Education and Welfare, working predominantly with their youth academy (eilte) players. In 2016, she was appointed to the role of Head of Safeguarding operating across all areas of the club.

Andy Sharpe, Head of Education and Welfare, Dipton Town FC Academy (English Football Championship).

Andy Sharpe has been Dipton Town's Head of Education and Welfare since 2014, a role that sits within the club's academy structure. Prior to this, he was a part-time educational tutor in the academy, having begun his career as a physical education teacher in further (post-16) education.

Felicity Archer, Head of Safeguarding, Highfield FC (English Premier League).

Felicity Archer has operated as Head of Safeguarding at Highfield FC since 2012, having previously fulfilled the role of Training Manager for the club's education provider. She began life at Highfield as a match-day steward which led to a coordination and training role with stewards at other local clubs.

Jon Clark, Head of Welfare, Tanfield FC (English Football Championship).

Jon Clark has been the Head of Welfare at Tanfield FC since 2018. Prior to this, he was involved in the club's academy as an educational tutor having begun his career as a college lecturer in sports science. A trained sports therapist, he also spent 12 years in a part-time role in the medical department of another local club.

Neil Cordingley, Head of Education and Welfare, Hobson FC (English Football Championship).

Neil Cordingley has been Head of Education and Welfare at Hobson FC since 2016, having previously been the Education Officer for the club's Community Department for five years. As a youngster, he played for Hobson's Centre of Excellence.

Owen White, Head of Safeguarding, Sherburn United FC (English Premier League).

Owen White has been the Head of Safeguarding at Sherburn FC since 2016. He previously served in the police force and has 20 years' experience in safeguarding. He has also been a part-time football coach at another Premier League academy.

5. Safeguarding and Elite Youth Football

As we have seen, the profile of safeguarding in English professional football has risen exponentially in recent years as a consequence of the pioneering work of Brackenridge et al. (2004, 2005). This has influenced a culture-shift within elite youth sport (National Society for the Prevention of Cruelty to Children) a development that Edith Ford of Sunnyside Rovers bore witness to:

Within football we've had the disclosures of the … historical sexual abuse, and that's, probably when I came into post. Looking back to 2004, society in general, people didn't talk about it. Prior to that there was an acceptance that these things were happening but nobody ever spoke about it. I guess what's changed now is people are not afraid to come [to a safeguarding officer] and they expect something to happen; it's that expectation that they come and there's a disclosure and that something will happen.

It is worth noting that none of the respondents in this study referred to allegations of sexual abuse within their own clubs and only broached this matter in order to emphasize the increasing profile and scope of safeguarding within the professional game. It is also worth noting that recent research into safeguarding has shifted from being focused solely on sexual abuse to cover a wider range of contextual issues (see Brackenridge and Rhind 2014). The current Child Protection in Sport (CPSU) Strategic Plan, developed in partnership between the NSPCC and Sport England, states that the focus for safeguarding in sport should be on elite athletes as they are amongst the most likely to face discrimination, exclusion and abuse due to the unique pressures associated with performance level sport (Sport Safeguarding Partnership 2016). Although some generic safeguarding issues such as underage drinking and social media usage emerged during interview discussions, the majority of the risks that arose during these conversations related to the specific environment in which elite youth footballers operate. Owen White of Sherburn United provided an example of the protective 'bubble' that elite footballers can become accustomed to and the subsequent welfare risks that may arise:

Yes, they [the young players] may be very wealthy but they have the same problems as you and I do and they potentially are more vulnerable than you and I because … they've been cosseted from lots of life's ups and downs. Yet they're going to face them at different stages of their life, and if they've never faced a crisis in a relationship and the first crisis they face, they happen to be in the peak of their career, who do they turn to?

Another wider safeguarding risk that was highlighted by respondents was how youth footballers occupied the considerable amount of down-time that their lifestyles afford. Jon Clark of Tanfield FC Academy shared his specific concerns over the gambling habits of players:

And I do worry constantly about the amount of time they have and if they're getting involved in things they shouldn't do because it can lead them into problems of addiction and things like that. … It's [gambling] something that gives them an adrenaline kick like they get on a match day. … I saw an Under-12 boy playing a football game, and one of the things he's got to do before he gets on the game is have a roulette spin to get the colours for his team. And I'm just thinking, what you're doing there quite cleverly, is getting the boys to enjoy the roulette experience, so you're drawing them in at 10, 11, 12 …

The most dominant safeguarding issue to emerge in interview was that of 'bullying', with four of the six respondents providing recent examples of both verbal and physical bullying that had taken place within their club. This was another issue around which Jon Clark (Tanfield FC) had concerns:

And there's the kind of welfare issues that go alongside … So, for example, we'll have accusations of bullying, from boys to boy, or from parents about another boy, accusations that maybe … staff could do a little bit more in supporting their boy.

As we have seen, allegations of bullying are not uncommon within the context of elite youth football amidst what has been identified as a hyper-masculine culture of power and control (Parker 1995; Pitchford et al. 2004; Brackenridge et al. 2004). Andy Sharpe of Dipton Town suggested that behaviours which could be interpreted as peer-group bullying were occasionally evident within his club:

Bullying's a strong word. But some of the boys saying some unkind things in training or so on … they're typical 'bread and butter' issues that come up from time to time for schoolboys.

Whether bullying is too strong a word to describe such relational dynamics is open to debate and, as we have seen, similar issues have been called into question within recent literature and policy surrounding the wider welfare of elite athletes (see Mountjoy et al. 2015; International Centre of Ethics in Sport 2016; Grey-Thompson 2017; Ong et al. 2018). In order to provide a safe and supportive environment in which players can thrive, Jon Clark went on to argue that a clear distinction needed to be made between 'banter' and bullying:

> And I also think that ... allegations made by parents about their sons being bullied have increased significantly and I think there's a lot of parental education needed in terms of what is an appropriate form of 'banter' and what's not.

Grey-Thompson (2017) corroborates the need to provide clarity in this area, suggesting that whilst banter is usually light-hearted, occasional and reciprocal, it can lead to bullying which is one-sided, intentional and persistent, unless clear boundaries are established and maintained (see Parker 2001, 2006). The potential consequences of failing to form this distinction were highlighted by Edith Ford, who explained that one player's experience of alleged 'targeted abuse' from a coach at Sunnyside Rovers led to them contemplating suicide, resulting in an internal investigation.

In order to combat these issues, clubs now provide educational workshops for their academy players around anti-bullying[7], as Andy Sharpe of Dipton Town FC explained:

> I'll do anti-bullying sessions so again, from time to time you'll get it from team mates, or a boy who's not playing well at the moment, we'll try and nip it in the bud ... educate them about ... the importance of all getting on; because that's going to make them a better team and play better individually.

These shifts in the profiling of safeguarding and the education of youth players should be seen as progressive. Yet, the motive to provide such training, often appeared to be based on optimal team performance rather than the holistic development of the athletes in question. It could be argued that the offer of additional resources in the shape of the pastorally-focused club chaplain could help contribute towards the wider welfare support of elite youth football players and the following discussion explores perceptions of club chaplaincy and the potential to utilize its services in this way.

6. Safeguarding, Chaplaincy and Welfare Support

Without exception respondents spoke positively about the role of the chaplain within their clubs in terms of contributing to the wider welfare support network of elite youth footballers. Central to their observations was the importance of the chaplain becoming embedded within the academy or wider club environment. This was most clearly articulated by Andy Sharpe of Dipton Town FC:

> He [the chaplain] comes in on a Thursday and he sort of makes his way around the club really and he's a familiar face. [He's] part of that support network; part of the Dipton Town family. All of our academy players are aware of the club chaplain. He also does an introductory talk as part of the 'Lifestyle Management' [curricular] programme with our Under 18s and our [Under] 23s.

Neil Cordingley also provided insight into how the chaplain at Hobson FC had developed relationships with the staff and players thereby reflecting the incarnational approach typically adopted by sports chaplains in the UK (see Mason 2006; Boyers 2016):

> Yeah, with being here for sort of seven years, in or around the club, obviously I've seen Bob [chaplain] quite a lot when he comes in on a Thursday ... And he comes around and has a

---

[7] Respondents also provided details of workshops on wider safeguarding issues such as consensual relationships, addiction and social media use.

chat, just checks if everyone's OK. But, y'know, there's genuine support there from him, that if anyone needs to speak to [him] . . . they can do.

Two respondents from Championship clubs stated that having a club chaplain was recognized as 'good practice' by League Football Education (the governing body of academy curricular) and by Ofsted (the UK government accountability body for curricular delivery). It could be argued that this is due to the chaplain focusing on the wider welfare of players through the provision of consistent pastoral support as opposed to performance issues (Roe and Parker 2016). In turn respondents provided examples of their club chaplains conducting religious ceremonies such as carol services, weddings and funerals as well as the pastoral care services that surrounded these significant life events.

That said, chaplaincy had clearly not always engendered such positive responses. For example, Edith Ford spoke candidly about how at one time, Sunnyside Rovers staff used to 'run for cover' when the club chaplain appeared. In their qualitative study of one English Premier League Academy, Roe and Parker (2016) argue that some players may be reluctant to utilise the chaplain due to their fears of their evangelistic agenda. Edith Ford went on to explain how she had come to view her club chaplain as faith-based, but not faith-biased. Likewise, Owen White shared his initial reservations about the chaplain at Sherburn United and how his thoughts had changed over time:

> Once I understood what Charlie's motivation was and that it wasn't to fill his congregation, I very quickly switched onto the fact that this was a free resource that I could tap into. His role as I see it is very much to be visible, to be there, to engage with people and to offer them, if they feel they need it, someone who they can talk to.

Owen's statement about chaplaincy as a 'free resource' paints a typical picture of how UK sports chaplains operate. None of those featured in this research were paid employees of their respective football clubs; rather, they acted as volunteers or within their role as local clergy (Weir 2016; Boyers 2016). A dominant trend to emerge from interview discussion was the significance of the independent nature of the chaplain and how this might positively impact working practice. Edith Ford provided a comprehensive analysis of this:

> If I'm unsure about something, I can chat [it] through with him [the chaplain] and he'll give me a really honest opinion and I know that that's not going anywhere and it's not coloured by any other football opinions . . . Because within a football club, everything is . . . it's not real. Everything is twisted and skewed because football is everything; football has got its own rules; they do what they want. Y'know, but actually there's rules and regulations out there. So sometimes . . . it's good to sort of step back and to speak to someone who's not involved with that . . . So, he is very, very neutral. And . . . for a lot of . . . things he doesn't have that emotional involvement. He's perhaps not as close [to the club]—sometimes I get a bit close—he's better placed.

As well as providing a sense of objectivity and impartiality, it could be argued that a further benefit of chaplains not being directly employed by clubs is the added layer of independence from management and coaching staff that this provides (Gamble et al. 2013; Roe and Parker 2016), something which Jon Clark reflected upon:

> I think sometimes the boys would be frightened to say [to a member of the management or coaching staff]: 'This has happened at the weekend, so I've been involved in this situation' as an example. 'Cos, if the club find out, they won't be happy. And they don't see Roland [the chaplain]—because he doesn't wear the tracksuit—as a direct employee of the club. They just see him as a nice guy to talk to.

Such a scenario is very much in line with Comfort (2006) view that players need to feel as though the club chaplain is representing them as opposed to representing the club itself. This is something that being separate and distinct from the management structures of the club and team hierarchies facilitates.

Addressing her fundamental belief that winning should not be to the detriment of player welfare, Grey-Thompson (2017) recommends that independent and confidential support is made available for athletes. By fulfilling this duty, chaplains enable elite youth footballers to discuss personal issues without harboring concerns regarding team selection, as expressed by Felicity Archer at Highfield FC:

> Well, they can say anything can't they, so they can say if they're down, or they can say if they're whatever—they can tell him [the chaplain] anything because he's not going to make a decision on whether he's [the player] going to be on the pitch on Sunday or not.

As well as the pressure to be selected every week, elite youth footballers also face the prospect of potentially having their contract terminated and being released from their academy at the end of the season. Owen White at Sherburn United discussed the longer-term implications of players having access to an independent, confidential source within this context:

> Someone who's not going to say to his coach "He can't play because he's got problems at home". Someone who's not going to stop him playing. Somebody who's not going to have influence on whether he stays in the academy or not. Somebody who's not going to think he's weak and perhaps . . . y'know, make decisions on him as an individual that's in any way, shape or form, connected to his football. [This is] Vital . . .

In terms of whether and to what extent chaplains should (continue to) be utilised in safeguarding provision, all respondents were extremely positive, providing several practical suggestions which included: being a consistent presence and an appropriate role model, and contributing towards positive mental wellbeing by encouraging players to speak up about their personal lives.

The demand for the continued input of chaplains was primarily driven by a need to increase resource and capacity around wider welfare support, with designated club staff often responsible for numerous players across multiple age groups. Unlike other organizational personnel, chaplains are able to focus their time and energy on the provision of pastoral support (Heskins 2006). In addition, given that the interests of academy staff are performance related their skill set is not geared towards pastoral and holistic support (Grey-Thompson 2017; Nesti 2010). An example of 'best practice' in response to these issues was provided by Sherburn United, who utilised a team of chaplains, with younger volunteers operating at academy level whilst the lead chaplain focused on senior players and club staff.

Another way in which such responsibilities might be devolved is through the intentional establishment of relationships between sports chaplains and other helping professionals. Indeed, recent research has suggested that there may well be a degree of overlap between the role of the sports chaplain and that of the sport psychologist, the former being viewed as someone who is primarily interested in the broader wellbeing of the individual, the latter as being primarily concerned with the enhancement of player performance (Nesti 2010). Gamble et al. (2013) observe that the sport psychologist and the sports chaplain bear similarities in their roles particularly concerning pastoral care, but that sports chaplains can also offer spiritual support which may facilitate discussion around specific faith issues as well as enhancing player contentment. Moreover, Nesti (2010) identifies an "overlap regarding issues of confidentiality and trust" (p. 109) across these roles to the extent that a cohesive working relationship between the two would appear sensible (Gamble et al. 2013). This is practically borne out by Hemmings and Chawner (2019) who provide a highly transparent account of the way in which such collaborative relationships might bear fruit. The findings presented here demonstrate that sports chaplains can indeed make a valuable contribution towards the safeguarding and wider welfare support of elite youth football players. In turn, greater intentionality by chaplains around the development of such relationships may well result in the provision of a more holistic support package for the athletes concerned.

7. Conclusions

The aim of this paper has been to investigate the extent to which sports chaplains might contribute meaningfully towards the safeguarding and wider welfare support of elite youth athletes. In particular, the paper has provided an indication of the kinds of issues that young players might face within the context of English professional football and how other helping professionals negotiate and facilitate holistic welfare support.

Data findings reveal a number of welfare concerns including vulnerabilities surrounding the use of peer-group 'banter' which may be interpreted as bullying. It was acknowledged that safeguarding provision has developed significantly in recent years, particularly within football, and that young players are encouraged to talk about their concerns. However, an emerging theme was the reluctance of elite youth players to share personal issues with coaching staff due to a fear of jeopardising their futures. As a consequence of their independence from team management, sports chaplains are able to focus on the personal needs of players thereby fulfilling the recommendation of safeguarding campaigners to adopt a child-centred approach to elite sport development. This unique position of impartiality led respondents to endorse the potential of sports chaplains to contribute (or contribute further) towards the safeguarding and wider welfare support of elite youth footballers. Yet despite this acknowledgement, respondents made no specific reference to any formal or structured safeguarding provision with which club chaplains were involved, a trend which is also noted in the extant literature. Although this position may reflect the typically marginalized nature of the chaplain's role, it could also mirror the view expressed by two respondents that their initial perception of chaplaincy presumed evangelistic motives. That said, both respondents expressed a modified understanding of the chaplaincy role having witnessed their work. It could be argued that the strongest advocates of chaplaincy were those respondents who emanated from non-footballing backgrounds and who appeared to express a greater understanding and appreciation of holistic pastoral care. Conversely, respondents with coaching or playing experience within the professional game highlighted the strengths of the chaplaincy provision in relation to meeting educational standards and enhancing team performance. This observation indicates the need to increase awareness of the wider benefits of chaplaincy in terms of wider welfare support.

These findings have implications for policy and practice both within sports chaplaincy and elite youth football. Firstly, sports chaplains should be encouraged to provide specialized services to football academies in order that players and staff within these settings might benefit from their wider pastoral support. In turn, sports chaplaincy organisations should be encouraged to establish specific training events concerning safeguarding provision. It is also recommended that sports chaplains should better promote these services to host clubs and governing bodies in order to emphasise their potential contribution to safeguarding. In addition, further research should be carried out on the role of sports chaplains alongside existing welfare provision in order to assess their potential to make a more strategic contribution to holistic athlete support.

This study focused on elite youth football players however, corresponding research could explore the contribution of sports chaplains towards safeguarding provision in female or amateur/grassroots football and/or across other sports. Alternatively, future investigations might concentrate on the chaplain's potential to contribute to specific elements of safeguarding such as the influence of social media on young players thereby reducing risks to athlete welfare.

Author Contributions: Oliver 75%, Parker 25%.

Funding: This research received no external funding.

Conflicts of Interest: The authors declare no conflict of interest.

References

Baker, Joseph, Stephen Cobley, Jörg Schorer, and Nick Wattie. 2017. *Routledge Handbook of Talent Identification and Development in Sport*. London: Routledge.

Baker, Math. 2006. You only sing when you're winning. In *Footballing Lives: As Seen by Chaplains in the Beautiful Game*. Edited by Jeffrey Heskins and Math Baker. Norwich: Canterbury Press, pp. 93–104.

Barker-Ruchti, Natalie, ed. 2019. *Athlete Learning in Elite Sport: A Cultural Framework*. London: Routledge.

BBC. 2018. Eighty Sports Coaches Convicted of Child Sex Abuse, Says Offside Trust. Available online: https://www.bbc.com/sport/football/46453955 (accessed on 22 August 2019).

Boyers, John. 2016. Sports chaplaincy in the United Kingdom. In *Sports Chaplaincy: Trends, Issues and Debates*. Edited by Andrew Parker, Nick J. Watson and John B. White. London: Routledge, pp. 20–31.

Brackenridge, Celia, and Daniel Rhind. 2014. Child Protection in Sport: Reflections on Thirty Years of Science and Activism. *Social Sciences*, 326–40. [CrossRef]

Brackenridge, Celia, Darren Bishop, Sybille Moussali, and James Tapp. 2005. The Characteristics of Sexual Abuse in Sport: A Multi-dimensional Scaling Analysis of Events described in Media Reports. *International Journal of Sport and Exercise Psychology* 6: 395–406. [CrossRef]

Brackenridge, Celia, Joy D. Bringer, Claudi Cockburn, Gareth Nutt, Andy Pitchford, Kate Russell, and Zofia Pawlaczek. 2004. The Football Association's Child Protection Research Project 2002–2006: Rationale, Design and First Year Results. *Managing Leisure* 9: 30–46. [CrossRef]

Bryman, Alan. 2015. *Social Research Methods*, 5th ed. London: Oxford.

Caperon, John, Andrew Todd, and James Walters, eds. 2018. *A Christian Theology of Chaplaincy*. London: Kingsley.

Charmaz, Kathy. 2000. Grounded theory: Objectivist and constructivist methods. In *Handbook of Qualitative Research*, 2nd ed. Edited by N. Denzin and Y. Lincoln. Thousand Oaks: Sage, pp. 509–35.

Charmaz, Kathy. 2014. *Constructing Grounded Theory*, 2nd ed. London: Sage.

Comfort, Alan. 2006. Shall we sing a song for you. In *Footballing Lives: As Seen by Chaplains in the Beautiful Game*. Edited by Jeffrey Heskins and Math Baker. Norwich: Canterbury Press, pp. 39–49.

Conn, David. 2018. Aston Villa Remove Kevin MacDonald from Coaching after New Bullying Claims. *The Guardian*. December 13. Available online: https://www.theguardian.com/football/2018/dec/13/aston-villa-remove-kevin-macdonald-coaching-new-bullying-claims-gareth-farrelly (accessed on 22 August 2019).

Gamble, Richard, Denise Hill, and Andrew Parker. 2013. Revs and Psychos': Role, Impact and Interaction of Sport Chaplains and Sport Psychologists within English Premiership Soccer. *The Sport Psychologist* 25: 249–64. [CrossRef]

Grey-Thompson, Tanni. 2017. Duty of Care in Sport, Independent Report to Government. Available online: http://www.sportsthinktank.com/uploads/final-copy---duty-of-care-report-2018.pdf (accessed on 22 August 2019).

Giulianotti, Richard, Fred Coalter, Holly Collison, and Simon Darnell. 2019. Rethinking Sportland: A New Research Agenda for the Sport for Development and Peace Sector. *Journal of Sport and Social Issues*. [CrossRef]

Hemmings, Brian, Nick J. Watson, and Andrew Parker. 2019. *Sport, Psychology and Christianity: Welfare, Performance and Consultancy*. London: Routledge.

Hemmings, Brian, Nick J. Watson, Andrew Parker, Damien Clement, Monna Arrivinen-Barrow, and Mark Oliver. 2019. Introduction. In *Sport, Psychology and Christianity: Welfare, Performance and Consultancy*. Edited by Brian Hemmings, Nick J. Watson and Andrew Parker. London: Routledge, pp. 1–9.

Hemmings, Brian, and David Chawner. 2019. The sport psychologist and the club chaplain supporting welfare: Reflections on five years of collaboration in English professional cricket. In *Sport, Psychology and Christianity: Welfare, Performance and Consultancy*. Edited by Brian Hemmings, Nick J. Watson and Andrew Parker. London: Routledge, pp. 42–53.

Heskins, Jeffrey. 2006. Who are you, who are you? In *Footballing Lives: As Seen by Chaplains in the Beautiful Game*. Edited by Jeffrey Heskins and Math Baker. Norwich: Canterbury Press, pp. 1–13.

Heskins, Jeffrey, and Math Baker, eds. 2006. *Footballing Lives: As Seen by Chaplains in the Beautiful Game*. Norwich: Canterbury Press.

International Centre of Ethics in Sport. 2016. Safeguarding Youth Sport. Available online: http://ec.europa.eu/assets/eac/sport/forum/docs/2ii1-pa2013-sys.pdf (accessed on 22 August 2019).

Kavanagh, Emma J. 2014. *The Dark Side of Sport: Athlete Narratives of Maltreatment in High Performance Environments*. Bournemouth: Bournemouth University.

Knight, Roger. 2006. You don't know what you're doing. In *Footballing Lives: As Seen by Chaplains in the Beautiful Game*. Edited by Jeffrey Heskins and Math Baker. Norwich: Canterbury Press, pp. 68–79.

Manley, Andrew, Martin Roderick, and Andrew Parker. 2016. Disciplinary mechanisms and the discourse of identity: The creation of 'silence' in an elite sports academy. *Culture and Organization* 22: 221–44. [CrossRef]

Mason, Phil. 2006. By far the greatest team. In *Footballing Lives: As Seen by Chaplains in the Beautiful Game*. Edited by Jeffrey Heskins and Math Baker. Norwich: Canterbury Press, pp. 39–49.

Morris, Steven. 2019. Football Coach Bob Higgins Guilty of 45 Counts of Indecent Assault. *The Guardian*. May 23. Available online: https://www.theguardian.com/uk-news/2019/may/23/football-coach-bob-higgins-guilty-of-45-counts-of-indecent-assault (accessed on 22 August 2019).

Mountjoy, Margo, Daniel Rhind, Anne Tiivas, and Michele Leglise. 2015. Safeguarding the child athlete in sport: A review, a framework and recommendations for the IOC youth athlete development model. *British Journal of Sports Medicine* 49: 883–86. [CrossRef] [PubMed]

National Society for the Prevention of Cruelty to Children. 2016. Available online: https://www.nspcc.org.uk/preventing-abuse/safeguarding/ (accessed on 22 August 2019).

Nesti, Mark. 2010. *Psychology in Football: Working with Elite and Professional Players*. London: Routledge.

Nesti, Mark, and Chris Sulley. 2015. *Youth Development in Football. Lessons from the World's Best Academies*. London: Routledge.

Ong, Chin Wei, Philippa McGregor, and Cherrie Daley. 2018. The Boy Behind the Bravado: Player advanced safety and support in a professional football academy setting. *Sport and Exercise Psychology Review* 14: 1–30.

Parker, Andrew, and Andrew Manley. 2016. Identity. In *Studying Football*. Edited by Ellis E. Cashmore and Kevin Dixon. London: Routledge, pp. 97–112.

Parker, Andrew, and Andrew Manley. 2017. Goffman, Identity and Organisational Control: Elite Sports Academies and Social Theory. *Sociology of Sport Journal* 34: 211–22. [CrossRef]

Parker, Andrew. 1995. Great Expectations: Grimness or Glamour? The Football Apprentice in the 1990s. *The Sports Historian* 15: 107–28. [CrossRef]

Parker, Andrew. 2001. Soccer, Servitude and Sub-Cultural Identity: Football Traineeship and Masculine Construction. *Soccer and Society* 2: 59–80. [CrossRef]

Parker, Andrew. 2006. Lifelong Leaning' to Labour: Apprenticeship, Masculinity and Communities of Practice. *British Educational Research Journal* 35: 687–701. [CrossRef]

Parker, Andrew. 2016. Staying Onside on the Inside; Men, masculinities and the research process. In *Gender Identity and Research Relationships*. Edited by Michael R. M. Ward. Studies in Qualitative Methodology. London: Emerald Group Publishing Limited, vol. 14, pp. 105–22.

Parker, Andrew, Nick J. Watson, and John B. White, eds. 2016. *Sports Chaplaincy: Trends, Issues and Debates*. London: Routledge.

Pattison, Steven. 2015. Situating Chaplaincy in the United Kingdom: The acceptable Face of 'Religion'? In *A Handbook of Chaplaincy Studies: Understanding Spiritual Care in Public Places*. Edited by Christopher Swift, Mark Cobb and Andrew Pattinson. London: Ashgate, pp. 13–30.

Pitchford, Andy, Celia Brackenridge, Joy D. Bringer, Claudi Cockburn, Gareth Nutt, Zofia Pawlaczek, and Kate Russell. 2004. Children in Football: Seen but not Heard. *Soccer in Society* 5: 43–60. [CrossRef]

Professional Footballers' Association. 2017. PFA Youth Advisory Booklet 2017–18. Available online: https://www.thepfa.com/-/media/Files/PFA-Youth-Advisory-Booklet-2017-2018.pdf (accessed on 22 August 2019).

Roan, Dan. 2019. Manchester City Launch Child Sexual Abuse Victim Payment Scheme. Available online: https://www.bbc.com/sport/football/47532719 (accessed on 22 August 2019).

Roderick, Martin. 2006. *The Work of Professional Football: A labour of Love?* London: Routledge.

Roe, Christopher, and Andrew Parker. 2016. Sport, Chaplaincy and Holistic Support (The Elite Player Performance Plan (EPPP) in English Professional Football. *Practical Theology* 9: 169–82. [CrossRef]

Slater, Victoria. 2015. *Chaplaincy Ministry and the Mission of the Church*. London: SCM Press.

Sport England. 2016. Towards an Active Nation—Strategy 2016–2021. Available online: https://www.sportengland.org/media/10629/sport-england-towards-an-active-nation.pdf (accessed on 22 August 2019).

Sport Safeguarding Partnership. 2016. Strategic Plan. Available online: https://thecpsu.org.uk/resource-library/2017/sport-safeguarding-partnership-2016-2021-strategic-plan/ (accessed on 22August 2019).

Strauss, Anslem, and Juliet Corbin. 1998. *Basics of Qualitative Research: Techniques and Procedures for Developing Grounded Theory*. Thousand Oaks, CA: Sage Publications, Inc.

Swift, Christopher, Mark Cobb, and Andrew Pattinson, eds. 2015. *A Handbook of Chaplaincy Studies: Understanding Spiritual Care in Public Places*. London: Ashgate.

Taylor, Daniel. 2018. Barry Bennell: The predatory Pied Piper who made stars and shattered lives. *The Guardian*. February 15. Available online: https://www.theguardian.com/football/2018/feb/15/barry-bennell-abuse-manchester-city-crewe (accessed on 22 August 2019).

The English Football League. 2018. Charter for Academy Players and Parents. Available online: https://www.efl.com/siteassets/efl-documents/charter-for-academy-players-and-parents-2017-18-e-book-1.pdf (accessed on 22 August 2019).

The Football Association. 2015. *Grassroots Football Safeguarding Children*. London: The Football Association.

The Football Association. 2017. Keeping Football Safe and Enjoyable—English Football's Ongoing Work to Protect Young People Who Play or Participate in Our National Sport. Available online: http://www.thefa.com/-/media/cfa/global/files/safeguarding/keeping-football-safe-enjoyable.ashx (accessed on 22 August 2019).

The Premier League. 2018. Elite Player Performance Plan. Available online: https://www.premierleague.com/youth/EPPP (accessed on 22 August 2019).

Wallace, S. 2019. Peter Beardsley Leaves Newcastle United Following Bullying Allegations. *Daily Telegraph*. March 6. Available online: https://www.telegraph.co.uk/football/2019/03/06/peter-beardsley-leavesnewcastle-united-followingbullying-allegations/ (accessed on 22 August 2019).

Weir, Stuart. 2016. Sports chaplaincy, a global overview. In *Sports Chaplaincy: Trends, Issues and Debates*. Edited by Andrew Parker, Nick J. Watson and John B. White. London: Routledge, pp. 9–19.

religions

MDPI

Article

Tracing the Landscape: Re-Enchantment, Play, and Spirituality in Parkour

Brett David Potter

Faculty of Humanities and Social Sciences, Sheridan College, Oakville, ON L6H 2L1, Canada;
brett.potter@mail.utoronto.ca

Received: 2 August 2019; Accepted: 26 August 2019; Published: 28 August 2019

check for
updates

Abstract: Parkour, along with "free-running", is a relatively new but increasingly ubiquitous sport with possibilities for new configurations of ecology and spirituality in global urban contexts. Parkour differs significantly from traditional sports in its use of existing urban topography including walls, fences, and rooftops as an obstacle course/playground to be creatively navigated. Both parkour and "free-running", in their haptic, intuitive exploration of the environment retrieve an enchanted notion of place with analogues in the religious language of pilgrimage. The parkour practitioner or *traceur/traceuse* exemplifies what Michael Atkinson terms "human reclamation"—a reclaiming of the body in space, and of the urban environment itself—which can be seen as a form of playful, creative spirituality based on "aligning the mind, body, and spirit within the environmental spaces at hand". This study will subsequently examine parkour at the intersection of spirituality, phenomenology, and ecology in three ways: (1) As a returning of sport to a more "enchanted" ecological consciousness through *poeisis* and touch; (2) a recovery of the lost "play-element" in sport (Huizinga); and (3) a recovery of the human body attuned to our evolutionary past.

Keywords: parkour; free-running; religion; pilgrimage; poiesis; ecology; urban

1. Introduction

Over the past two decades, the sport known as parkour has become a global phenomenon, with groups of practitioners or *traceurs* emerging from Paris to Singapore. Recent world championship competitions have been held in Turkey, Brazil, and the U.S.A., featuring athletes from locations as varied as Russia, Greece, Portugal, France, Germany, and England. There is even an active parkour squad in Gaza, practicing their art in buildings destroyed by armed conflict.[1] Jeffrey Kidder has written of the "global ethnoscape" of parkour, which captures something of the ethos of this subculture; *traceurs* and *traceuses* from around the world have become connected via YouTube and other social media, sharing videos of difficult techniques and innovative movements, creating a global community which inhabits a kind of boundless, virtual meta-city (Kidder 2017, p. 48).

Although parkour and its close cousin "free-running" are relatively new phenomena, they have already prompted theorists to conceptualize a range of new approaches to human mobility and the ontology of the globalized city. Both parkour and free-running differ significantly from traditional sports in their use of existing urban topography including walls, fences, stairwells, rooftops as an obstacle course/playground to be creatively navigated. There are also lively debates as to whether parkour can be classified as a sport, as it actively discourages competition (Bardwell 2010, p. 22); instead, parkour in its purest form purports to simply be about the ability to move efficiently through a given environment.

---

[1] See https://www.youtube.com/watch?v=8Q8rYz3T1-8.

Alongside discussions of the political and social dimensions of space and movement, are there resources from the academic study of religion that could be brought to bear on the study of parkour? In his ethnographic study, Michael Atkinson comes to define parkour as "a mode of bringing forth or revealing dimensions of the physical and spiritual self through a particular type of urban gymnastics" (Atkinson 2009, p. 169). In its haptic, intuitive exploration of the urban environment, it is not fanciful to suggest that parkour not only undertakes a kind of creative refashioning of the individual's relationship to the cityscape, but involves the city itself in a process of *poeisis*, as several theorists have described it, full of spiritual meaning. Such an approach can be contrasted to the "disenchanted" landscape associated with the rise of modern capitalism; parkour provides an alternative mode of engagement with spaces shaped by the grid-like imaginaries of late modern capitalism, by rehabilitating an "enchanted notion of place which, through wonderment, imagination and participation, is in continuous composition" (Saville 2008, p. 892). In Atkinson's terminology, the parkour practitioner or *traceur/traceuse* is interested in "human reclamation"—a reclaiming of the body in space, and of the cityscape itself—which can be seen as a form of playful, creative spirituality based on "aligning the mind, body, and spirit within the environmental spaces at hand" (Atkinson 2009, p. 169).

This study will examine this aligning, reclaiming, and "re-enchanting" aspect of parkour and free-running at the intersection of spirituality and ecology in three ways. First, it will be examined as a returning of sport to a more enchanted ecological consciousness through *poeisis* and touch, in connection with a notion of the *traceur* as a kind of urban pilgrim. Secondly, parkour is posited as a recovery of the lost "play-element" in sport, drawing on the work of Johann Huizinga. Finally, with the first two approaches in mind, the sport is investigated as a recovery of the human body in its natural environment attuned to our deep evolutionary past.

Though still comparatively new, the scholarly literature on parkour has ably situated it in reference to the organization of space in capitalist societies; transnational politics; constructions of the virtual; as well as in relation to Csikszentmihalyi's notion of "flow" (for an overview, see Csikszentmihalyi 1990, pp. 6–11; Kidder 2017, p. 57). This study hopes to provide new theoretical avenues for considering parkour by introducing terms and models from the study of religion. It is hoped that such crossdisciplinary exchanges will aid, rather than hinder, further contextual study of the social roles of parkour in the global city.

2. History and Core Values

There are unquestionably quasi-religious values which have accompanied the sport of *le parkour* ever since it was first developed in the 1990s by David Belle in France. It is sometimes described as *l'art du déplacement*—a phrase which helps underline its simultaneous existence as a sport, an art, and a rigorous discipline. At times the movement can resemble a religious order. For example, reading the philosophical "Charta" of the parkourONE group, published in 2014 by an influential contingent of traceurs in Germany and Switzerland, one encounters doctrinal statements which clearly promote a particular version of orthodoxy, establishing the squad within a particular interpretive tradition. With small modifications, such a manifesto could just as easily describe a restorationist religious ideology:

> "Thus, we perceive ourselves as bearer of a specific idea; Parkour cannot be redefined randomly but rather has its own (even conceptual) history. We distinctly declare: David Belle is a living person; as such he and his perception, likewise, will change. We follow David's original idea, however, not him as a "Guru", leader or the like".[2]

For the parkourONE group, it is the core ideas of parkour, not Belle himself, which are at the center of their practice. The five "fingers" or guiding principles listed on the parkourONE Charta, which correspond to the five pillars of parkour as generally accepted by the global community, are

---

[2]   Available from https://parkourone.com/wp-content/uploads/2014/03/20160305_Charta_english.pdf.

non-competition, caution, respect, trust, and modesty. Such values, along with slogans such as "être et durer" (to be and to live on) and "être fort pour être utile" (being strong to be useful), are ethical norms which bind the parkour community together. These slogans also go back to the earliest roots of the sport in Georges Hébert (1875–1957) and his *méthode naturelle*, which is frequently cited as a precursor to parkour. "Hébertisme" emerged after the First World War in France as a new mode of physical exercise to train soldiers. In Hébert's system, individuals who practiced the Natural Method, consisting of running, jumping, lifting, climbing, and balancing in ways that mirrored these movements as one might undertake them in simply navigating the natural environment, would develop not only physically but morally (see Atkinson 2009, pp. 171–72). The mantra of "être fort pour être utile" was, for Hébert, a way of reconnecting exercise to lived experience, embedded in nature and daily life. In his view, influenced by his experience abroad in the military, a return to a more natural method of physical activity was a return to an idealized past condition, one he romantically (and problematically) believed to be the way people still lived in Africa (Marshall 2010, p. 167):

> "The man in the state of nature, forced to lead an active life to support himself, realizes a full physical development by doing only useful and natural exercises and executing the most common physical labor". (Hébert 1912, p. 8)

Instead, for those in cities, this state of nature had been lost through the sin of sedentary existence:

> "In civilized countries, social obligations, conventions and prejudices move man away from the natural life outdoors and often prevent the exercise of his activity. His physical development is slowed or halted by these obligations or conventions". (Hébert 1912, p. 8)

Hébert's insights were taken up by the French army, and the style of exercise he developed was used widely in post-WWI France. It was here that Raymond Belle, David Belle's father, first trained in *parcours de combattant*, the obstacle-course fitness program that his son would eventually adapt for urban contexts.

Alongside this focus on the ethical path and the implicit valuing of a "way of nature", another religiously-tinged moment in the history of parkour follows the narrative of David Belle himself, who founded the sport in the early 1990s. If Belle is the prophet of this new method, originating *le parkour* in its purest form while self-consciously situating it in reference to the story of his father, his follower, Sébastien Foucan, goes on to found another variant sect ("free-running")—a classic story of divergence and succession with analogues with the formation of any religious denomination. Belle was worried about the "commercialization of parkour" (Bardwell 2010, p. 23), and believed that any creative expressions—choices of movement made for purely aesthetic reasons, such as impressive backflips and gymnastics influenced by *capoweira* or hip hop—are contrary to the spirit of parkour which must be about moving between points as efficiently as possible. This Great Schism is a foundational moment in the spiritual history of parkour and is frequently returned to in its texts.

Sébastien Foucan, who through his involvement with the Yamakasi squad subsequently helped develop "free-running" as a parallel, but more syncretic/synthetic blend of styles of movement, also speaks in more explicitly spiritual terms about the discipline:

> There is a link between everything, like in [the teachings of] martial arts, it's about energy and how we are all connected. Everything we do affects other people negatively or positively. I say, 'your life is a road, your feelings a guide, your body a vehicle'". [...] "Other people's journeys can influence my own journey–this is my teaching". (El-hage 2011, para. 7)

This moral pursuit of connection and energy through the beauty of free-running takes on ascetic dimensions, not unlike the stories of Mahavira from the Jain tradition; there are stories that the Yamakasi members would train without food and water and sleep without blankets in order to learn how to endure.

3. Parkour and Pilgrimage

In one of the seminal videos of early parkour, the trailer *Rush Hour* (2002), David Belle sits at a stereotypical desk, working at what we presume is a typical 9-to-5 job. Suddenly, he stands up, and removes his shirt, then steps out the window to gracefully assume the position of a handstand on the edge of the building. It is a moment of enlightenment and liberation, breaking free from the ordinary. He begins his journey over the rooftops of Paris, and his seamless, seemingly unearthly motion off of walls and between buildings is intercut with images of the traffic jam on the street below. Unconcerned with the activities of mere mortals, Belle has visually become an enlightened being who has transcended the "grid" of the 9-to-5 world–the world of traffic, busyness, and ugliness—and instead been raised to a new consciousness.

Such a comparison underlines the roots of the spirituality of parkour in both Eastern and Western paradigms. In writings on the subject, where parkour and free-running are defined as the "art of movement", the overcoming of obstacles in the "most quick, efficient, and flowing way" is seen to entail not only agility but "prudence, awareness, control, and cool-headedness" (Bardwell 2010, p. 23). Such sapiential qualities, in line with Stoic self-control, the Buddhist middle way, or the Daoist way of nature seem more like virtues than purely athletic skills.

> "A lot of different religions have a word for this. [. . .] The Hindus call it moksha. The Zen Buddhists call it a state of Zen awareness. [. . .] When you are engaged in parkour [. . .] you are envisioning your route. Everything else goes quiet for a moment. You feel the wind on your face. You feel the sweat dripping down your body. You feel muscles in your body moving, and in that moment, you are completely aware of your environment". (a young traceur named 'Eric', quoted by Kidder 2017, p. 57)

This experience of total awareness of what is going on in the present is notably similar to *Vipassana* or mindfulness meditation, where each experience is deeply felt as it rises and passes—a simultaneous privileging of immediacy and of the transience of the moment. Theorists have highlighted the distinctively "serene ethos" of parkour, which Mould connects to Eastern philosophies, as a kind of "passive rediscovery" of the world (Mould 2009, p. 740). This rediscovery is both meditative and creative; the attentive eye of the traceur/traceuse looks for paths through the city, and through tactile, kinetic engagement with its surfaces (concrete, wood, metal) transforms the complex geography of each city block into something which can be moved through with ease. A "Kong Vault", for example, transforms any obstacle into a kind of ad hoc touchdown point for a body moving at high speed, the arms thrusting the body forward in space; walls are mounted easily using "cat leaps" and "wall runs", each relying on momentum, precision, and fluidity of movement. Mould describes the way that Belle's movement "reappropriates the urban built environment from a striated space to a more fluid smooth space" (Mould 2009, p. 741), a kind of "gliding" through the solid textures of the city which reconfigures it into something new. The goal of the traceur within this urban ecology is to learn technique so well that individual movements are gradually integrated into constant motion—a smoothing out of form and a seamless integration of body and built environment.

This smoothing out through physical engagement is at the core of parkour as a journey (*parcours*) of re-enchantment. Many efforts have been made to link the "renegotiation of the environment" (Marshall 2010, p. 165) accomplished by parkour to its most obvious historical and conceptual predecessors: The Situationist *dérive* and the Benjaminian *flâneur* (Marshall 2010, p. 166; Atkinson 2009, p. 174). Yet another literary figure also comes to mind, one with a more explicitly religious meaning: The pilgrim, whose wanderings through space are oriented by a higher plane of consciousness.

What is a pilgrim? Most iterations in the major world religions are focused on travel to a particular sacred space, marked off for this purpose: The Ka'aba in Mecca, the cathedral of Santiago de Compostela on the edge of Spain, or the confluence of three rivers during the Kumbh Mela festival in India. Yet there is also a strong tradition of pilgrimage as peregrination—"wandering" without a particular destination in mind. Here one might draw on the "white martyrdom" of the Irish saints, the itinerant lifestyle of a mendicant friar, or the Tendai Buddhist monks of Mt. Hiei (the so-called "marathon monks") who prayerfully circumambulate a forest path around the mountain which takes years to traverse. In both cases, the wandering pilgrim is a figure on the margins, or in Victor and Edith Turner's terminology in a "liminal" space, freed from "profane structures" and instead immersed into the *communitas* of fellow wanderers (Turner and Turner 1978, p. 9). The pilgrim can only re-integrate into society after a symbolic transformation, which in turn transforms them into a living icon of the whole "redemptive tradition" (Turner and Turner 1978, p. 10). In other words, pilgrimage is a deeply embodied mode of spirituality, wherein the body becomes, through engagement with the landscape, a site of religious *poeisis* or meaning-making.

In an important essay, sociologist Zygmunt Bauman helpfully distinguishes between the "pilgrim" and the "tourist". While the tourist perceives only "fragmentariness and discontinuity, narrowness of focus and purpose, shallowness of contact" in relation to space, the itinerary of a pilgrim is a kind of deep experience of place, invested with spiritual meaning on each step of the journey.

> "Pilgrims had a stake in solidity of the world they walked; in a kind of world in which one can tell life as a continuous story, a 'sense-making' story, such a story as makes each event the effect of the event before and the cause of the event after, each age a station on the road pointing towards fulfilment". (Bauman 1996, p. 22)

For Bauman, the postmodern subject is a tourist, without a stable identity or sense of a destination—what exists instead is "life-as-strolling" (not unlike the *flâneur*) concerned only with the superficial aspects of the environment. Yet as Ivo Jer Asek points out, bringing the tourist/pilgrim discussion into conversation with the study of sport, there is a continuum of experience from "tourism to pilgrimage" and thus also of profane and sacred (Asek 2011, p. 223). Tourists may become pilgrims. What distinguishes between the two is not that they both aim to "visit and see" but that only the pilgrim continues on to "inspiration, feeling and deep experiencing" (Asek 2011, p. 231).

Within urban spaces disenchanted by social and economic distensions, parkour and free-running help "re-enchant" the landscape through pure, intense movement: in "deep experience" of urban topography, a kind of free-form peregrination over, under, and around walls, rooftops, and fences. Through physical exertion, the city itself is transformed into a pilgrimage "route" (*parcours*), circuitous and deeply intuitive, where each obstacle touched is a kind of sacred waypost. Perhaps the model of the traceur as postmodern pilgrim helps support the claim that in parkour one experiences "landscape and body as contingent, dynamic entities whose entanglement continuously blurs and (re)assembles body-landscape boundaries" (Bin Loo and Bunnell 2018, p. 148). The peregrinations of the traceur map out levels of meaning on the city which are both "symbolic and material", not unlike the mantras of a Buddhist monk or a medieval Christian pilgrim. Both pilgrimage, broadly considered, and parkour inhabit liminal social spaces where new ways of movement are superimposed onto the landscape—a process of creative mythmaking or *poeisis*, undertaken within the *communitas* of a subaltern community of athletes. Attention to this creative, communal aspect of the sport is crucial to understanding its connection to spirituality.

4. Parkour and Play

A second approach to the question of how the vocabulary of the study of religion can inform understanding of parkour and free-running is the concept of "play". Here it is important not to lose sight of the bigger picture. Paula Geyh has argued persuasively that an over-focus on the individual elements of parkour (particular jumps, flips, and vaults, etc.) leads away from what she terms a "poetics" of the sport: "the ways in which parkour can be seen to 'remap' urban space and to demonstrate a resistance to its disciplinary functions" (Geyh 2006, para. 5). By liquefying the enforced space of the grid, parkour is engaged in "creating a parallel, 'ludic' city, a city of movement and free play within and against the city of obstacles and inhibitions" (Geyh 2006, para. 10). Mould similarly calls into question the common notion of parkour as inherently subversive—seeking to "corrupt" urban spaces, perhaps like graffiti or trespassing–and instead points out its creative embrace of the city as "playground" (Mould 2009, p. 743; Ameel and Tani 2012, p. 18).

One of the most influential discussions of "play" is the now-classic *Homo Ludens* by Johann Huizinga. Huizinga describes the "play-element" as a formative aspect of human culture. Although he seems more interested in games than in physically demanding sports, his insights have direct relevance as they seek to demarcate the boundary between the ordinary and what is specially marked out for "play". For Huizinga, play is closely related to the imagination, and to the arts—it is a species of "significant form" not unlike a painting or a theatrical performance (a "play" in the proper sense), which imbues a particular constellation of actions and spaces with meaning.

Huizinga's description of play can be summarized under a few headings. First, play is a voluntary activity, marked by freedom; one cannot be compelled to play, and this provisional nature means that the game can be stopped at any time (Huizinga 1949, p. 8). Second, and here again the analogy with the arts comes to the fore, play exhibits a kind of relative "disinterestedness" in that it does not directly satisfy individual or social wants and desires (Huizinga 1949, p. 9). The game, with its "pretend" rules and suspension of ordinary life, is autotelic and self-sufficient; whatever desires are fulfilled are done so obliquely. Third, play satisfies "communal and individual ideals":

> "[Play] adorns life, amplifies it and is to that extent a necessity both for the individual–as a life function–and for society by reason of the meaning it contains, its significance, its expressive value, its spiritual associations . . . ". (Huizinga 1949, p. 11)

For this reason Huizinga also emphasizes play as "symbolical actualization"—not just a trivial activity, but like a mating dance or a religious ritual a solemn event of representation whereby "something invisible and inactual takes beautiful, actual, holy form" (Huizinga 1949, p. 14–15).[3] He thus draws attention to the inside/outside dichotomy which is instantiated in the game—the inner "circle" of the game, which spatially could entail the football field, the sumo ring, the golf course–is all that matters. The players, like priests or cosplayers, are set apart to form a social "circle" as well, there are players, and there are spectators, and this sacred line cannot be broached.

Huizinga's approach, then, draws significantly on the idea of demarcation—the way play is secluded or "limited" from ordinary/profane life. Both space and time are implicated in this seclusion. The game occurs within a limited time-frame: It is "played out" within a particular duration of time set aside for this purpose (whether a few hours for a basketball game, or several days for a cricket test match). It also occurs in a (de)limited space—a playing-ground marked off beforehand in some way. Within this spatial-temporal nexus, order is created: A "temporary, limited perfection" with its own aesthetic qualities which must not be deviated from (e.g., by a failure to follow the rules or by external events impinging on the performance of "the game")—it has its own code of law, manifesting the triumph of order over chaos, which translates in religious terms to holiness. Huizinga thus concludes, "formally speaking, there is no distinction whatever between marking out a space for a sacred purpose and marking it out for the purposes of sheer play" (Huizinga 1949, p. 20). Religious ritual is itself a form of play, both equally "serious" and "pointless" (Huizinga 1949, p. 19)–and so culture itself is formed at its deepest level by the ludic or "play-element".

How can this definition help inform study of parkour, particularly with reference to emergent concepts of spirituality? If play is the basis of ritual, parkour is itself a kind of ritual enactment of the very roots of culture. Michael Novak captures some of the spiritual significance of play within culture, suggesting that it upends traditional understandings of how the capitalist enterprise is related to the domain of spirituality:

---

[3] Here Huizinga is careful to link "representation" not just to *mimesis*, but to *methexis* or participation–where the ritual is not "merely imitative" but is "actually reproduced".

"Play, not work, is the end of life. To participate in the rites of play is to dwell in the Kingdom of Ends. To participate in work, career, and the making of history is to labor in the Kingdom of Means ... In a Protestant culture, as in Marxist cultures, work is serious, important, adult. Its essential insignificance is overlooked. Work, of course, must be done. However, we should be wise enough to distinguish necessity from reality. Play is reality. Work is diversion and escape". (Novak 1993, p. 170)

Although parkour and free-running are disciplines, as in the *Rush Hour* video they both go against the grain of the capitalist geography of "work" in their playful gliding through the grid-space. Play is construed as re-creation: The shaping of the urban environment through haptic poeisis. In Huizinga's terms, play itself becomes a kind of symbolic actualization, a meaning-making activity that transmutes one form into another.

What is most striking about Huizinga's approach to play is its emphasis on setting aside a dedicated time and space to the play-activity. Returning briefly to his underlying image of the work of art, we might say that this model echoes the modernist understanding of the painting as the autonomous, self-possessed phenomenon which governs the terms by which it may be beheld–the "sacred" space of the art gallery set apart for this ritual purpose. This model of aesthetic experience has been called into question both for the way it characterizes art as a solitary, abstracted experience but also because it conceives of art in only one particular way—*l'art pour l'art*, removed from any real-world concerns. Movements such as relational aesthetics, or Hal Foster's championing of the anti-aesthetic, are an attempt to think of art in non-modernist terms where its communal, embodied, and cultural aspects are not siphoned off but form precisely the hermeneutical matrix within which art makes meaning.

So too "play" can be considered in a non-autonomous, less hermetically isolated way. Any tendency to separate out the sacred and the secular, or for that matter the privileged domain of the "aesthetic" over against the ordinary, is precisely what is called into question by emergent practices such as parkour and free-running. Instead of occurring within a stadium, or even on a field dedicated to this purpose, "play" happens in and on the urban environment itself. For this reason, we might think of parkour's reconfiguration of space as essentially secular—although it might rely on an "enchanted" paradigm of the city, it is not confined to (culturally) "sacred" spaces like the stadium, gallery, or church. This is not to say that Huizinga's thought is itself focused only on the large scenario of the baseball stadium. By sacred space, he may mean something as small as the implicit and invisible boundaries of a children's game. What parkour seems to do, however, is open up the closed "circle" of the game to encompass the whole city.

This "secularization" of play, and its implications for spirituality, is mirrored by other cultural developments around sport which similarly transform the city into a site of play. For example, the charity marathons which regularly shut down highways and major thoroughfares in North American cities are another example of sport breaking the boundaries of dedicated space in order to create a sense of the embeddedness of sport in "ordinary" topologies. Similarly, augmented-reality video games such as Pokémon GO and Ingress evince a kind of transformation of the world into a game-space, and a democratization (though still technologically moderated) of who gets to be a "player". Though the inside/outside divide identified by Huizinga seems to persist—one is either "in" the game, immersed in a virtual world, or "out" in the real world—what is flattened out is the sense of the play-space as ritually distinguished from the profane.

Fragoso and Reis, in their research on location-based or "locative games" (Fragoso and Reis 2016, p. 132), invoke Huizinga's account of games and play in order to speak of what they term "ludic re-enchantment" wherever the urban environment is appropriated for play. What they have in mind are role-playing games where the player inhabits a fictional universe anchored in the "materialities of ordinary life" (135), but this term is equally applicable to parkour and free-running. Unlike the video game players cited by Fragoso and Reis 2016, however, the parkour athlete does not re-enchant space through social relationships or a fictional narrative; instead, parkour creates meaning through movement, in creative, adaptive, haptic encounter with the textures of the landscape. Poeisis, in this context, becomes not a matter of "simultaneous experience of fictional and real spaces" as in location-based gaming, but rather emerges out of the embodied experience of reconfiguring space. There is no alternate reality; rather, the physical world is transmuted into a "route", full of potential and spiritual energy–what matters is complete awareness and immediacy, a physically embodied ritual of sense-making. The whole city becomes a secular venue for such "ludic re-enchantment".

This shift towards a more secular iteration of play may simply echo institutional changes underway in society. Like religion and the arts, sport and play more generally is undergoing its own cultural shift out of traditional times and spaces to something less "holy". The unholiness of parkour extends this tendency in multiple ways. The special vestments which distinguish athletes from laypeople are gone; instead, the dress called for is ordinary, a kind of ascetic purity. There are no dedicated spaces for parkour, aside from the occasional sparsely furnished training "gym"-instead, as with other urban sports like skateboarding and even basketball, existing elements of urban architecture must be repurposed to become ramps and obstacles.

This does not necessarily mean that parkour and free-running represent a more pure or primitive form of athleticism. This is at stake in the debate between the more "purist" practitioners of the Natural Method and the gradual commercial adoption of free-running as a more performative, less "practical" style. The reasons for this are ideological. For the Belle school, parkour is resolutely focused on moving from one point to another in the most efficient way possible, displacing the body without attention to style. On the other hand, disciples of Foucan maintain that style and gymnastic flourishes are germane to the sport. Yet internationally, it seems that these two styles have re-integrated, so that parkour is essentially a blend of the two–a form of free, uninhibited play which is broadly consistent with Huizinga's quasi-religious phenomenology of play as "symbolic actualization".

5. Re-Enchantment and Human Evolution

One final core doctrine of the parkour ethos is built on a particular understanding of the history of the human body. Reading manifestos by practitioners, one frequently encounters the idea that exercise as practiced in contemporary Western society is unnatural in its repetitive, artificial techniques and subsequently has deleterious effects on the body. Activities focused on developing particular muscles in isolation, such as lifting weights, jogging on a treadmill, and other technologies for personal fitness are characterized as "a very recent phenomenon in human evolution". Traceur Dan Edwardes suggests in an article for the UK-based parkour training school PKGen that movement for fitness should be natural, adaptive, and variable, more akin to the way humans have naturally moved through environments for millennia. In other words, rather than consigning us to the artificial rhythms of the gym, physical exertion ought to connect us to a more "human" way of moving, "which from an evolutionary standpoint means covering terrain and getting over obstacles regularly" (Edwardes 2018). Parkour is thus construed as retrieving an older, more organic way of moving and exercising than any modern program of weights and machinery.

It is undeniable that jumping, vaulting, and especially running—particularly over long distances—are capabilities with origins deep in humanity's evolutionary past. Although one might be tempted to compare the agility on display in parkour and free-running to the movements of animals, including our simian cousins, there are ways to speak of a distinctively "human way of moving". In particular, a prehistoric account might trace these movements back to the emergent "striding bipedalism" of the genus *Homo*, which gives rise to fast running speeds over long distances unlike those of any other primates (Bramble and Lieberman 2004). The ability to run quickly may stretch as far back as *Homo australopithecus*, who lived between approximately 2–4 million years ago. Such endurance running may be one of the primary shaping factors of the modern human form, explaining the physiological divergence of modern humans from earlier ancestors. Moreover, since running played a role in the ability of prehistoric hominids to chase down prey, it may have been one of the most important factors which allowed humans to find an ecological niche and eventually grow to become the dominant species on the planet (Carrier et al. 1984).

"Free running" is thus perhaps at the very core of what it means to belong to the human race, both physiologically and panhistorically. Although it is difficult to substantiate the claim that the fluid, gymnastic movements of parkour approximate those undertaken by our earliest ancestors, and we ought to be wary of the colonial primitivism of the *méthode naturelle* from which parkour is derived, clearly there is for traceurs a connection between rapid movement and the primal experience of the human being. Yet although moving between points as quickly and efficiently as possible, the stated aim of parkour, may emerge out of evolutionary necessity, it paradoxically becomes a modality of "ludic re-enchantment" in the late modern landscape. To run freely, smoothly, and playfully through the urban environment is to in a sense retrieve an earlier mode of "deep experience", situating oneself in an ancient confluence of physical exertion and aesthetic sensation. In Atkinson's terminology, parkour involves a kind of re-alignment of body and environment, the physical and the spiritual. Moving across terrain and over obstacles becomes not 'just' a game, still less just a fitness program, but (again, with a nod to Huizinga) participation in an essentially human ritual action.

Of critical importance in terms of re-enchantment is the role of the community of traceurs/traceuses in transforming space together. It is the shared experience of movement and beauty, rather than the individual's connection to the environment alone, that invests each *parcours* through the city with symbolic meaning. It is together, whether locally or within the larger context of a virtual meta-city, that traceurs transform the secular spaces of the "grid" into new, fecund pilgrimages.

6. Conclusions

A constructive way forward for understanding the relationship between parkour and spirituality might focus on ecology. In Bauman's essay on the tourist and the pilgrim, he notes that we now inhabit a world "inhospitable to pilgrims" (Bauman 1996, p. 23) where the need to build identity (particularly to leave "traces" behind) leads to the barrenness of a figurative desert. The process of identity-formation, so key to the figure of the pilgrim, has become obsolete. Perhaps parkour, with its emphasis on embodied experience of the landscape, can help transform this desert into an oasis. An "enchanted" view of not only nature, but the urban landscape itself as full of potential for beauty and grace, could perhaps be the basis for a renewed ecological consciousness. This is certainly at the core of Atkinson's approach, who draws a comparison between parkour and the tradition of environmental transcendentalism to be found in Thoreau (Atkinson 2009, pp. 175–76). Parkour thus becomes "the art of revealing or bringing forth possibilities of the alternatively environmental self/society" (Atkinson 2009, p. 178). In other words, the trope of the city "smoothed out" by creative movement is bound up with a return to nature, particularly in our own era of climate change and the degradation of urban spaces. To give just one example, cities such as Toronto and Chicago are facing major problems related to flooding because the concrete out of which cities are built cannot absorb water like porous grass or forest land. Tracing a new path for water in the topology of the built environment is a kind of cultural analogue for parkour and free-running: Moving over and around obstacles efficiently and even gracefully.

Atkinson's ecologically-inflected model of the spirituality of parkour draws on the idea of "reclamation", and indeed this is an illuminating term. For the body itself, the roots of the sport in the *méthode naturelle* seem to ground the types of movements developed by parkour in the ancient history of humanity—a prehistory of human bodies before the rise of cities. In Huizinga's terminology, parkour brings to the foreground the ancient ludic element to human society, reminding us that all cultural activity–including religion–is suffused with "play". Though there is an aspect of romanticization here, it seems clear that a plurality of ways of moving and a recovery of play may indeed activate older, more primal modes of human mobility which need to be adapted within the new context of global cities. If there is a spirituality that lends itself to ecology, surely it is one which retrieves the embedded, embodied, playful nature of the human body in kinetic, tactile connection with its environment. Here the core parkour values of non-competition, caution, respect, trust, and humility can be understood as ecological imperatives as well; "être et durer" and "être fort pour être utile" are both individual virtues and spiritual values for the good of humanity.

This brief survey of how concepts of religion and spirituality can be used to explore the possibilities of body and environment in parkour suggests, then, that there are new paths to "trace" in considering the intersection of lived urban experience and the sacred/secular binary. In particular, there are ample reasons to "read" this emergent art/sport/discipline as a resource for ecologically-minded forms of spirituality which reclaim the human body and the city itself for symbolic, ludic ends.

Funding: This research received no external funding.

Conflicts of Interest: The author declares no conflict of interest.

References

Ameel, Lieven, and Sirpa Tani. 2012. Parkour: Creating Loose Spaces? *Geografiska Annaler: Series B, Human Geography* 94: 17–30. [CrossRef]

Asek, Ivo Jir. 2011. Pilgrimage as a Form of Physical and Movement Spirituality. In *Theology, Ethics, and Transcendence in Sports*. New York: Routledge, p. 223ff.

Atkinson, Michael. 2009. Parkour, Anarcho-Environmentalism, and Poesis. *Journal of Sport & Social Issues* 33: 169–94.

Bardwell, Jeff. 2010. Parkour: The Nature of Sport and its Ethical Possibilities. In *Philosophy of Sport: International Perspectives*. Edited by Alun Hardman and Carwyn Jones. Newcastle: Cambridge Scholars Publishing, pp. 21–37.

Bauman, Zygmunt. 1996. From Pilgrim to Tourist: Or a Short History of Identity. In *Questions of Cultural Identity*. Edited by Stuart Hall and Paul Du Gay. London: SAGE Publications, pp. 18–36.

Bin Loo, Wen, and Tim Bunnell. 2018. Landscaping Selves Through Parkour: Reinterpreting the Urban Environment of Singapore. *Space and Culture* 21: 145–58.

Bramble, Dennis M., and Daniel E. Lieberman. 2004. Endurance running and the evolition of *Homo. Nature* 432: 345–52. [CrossRef] [PubMed]

Carrier, David R., A. K. Kapoor, Tasuku Kimura, Martin K. Nickels, Satwanti, Eugenie C. Scott, Joseph K. So, and Erik Trinkaus. 1984. The energetic paradox of human running and hominid evolution. *Current Anthropology* 25: 483–95. [CrossRef]

Csikszentmihalyi, Mihaly. 1990. *Flow: The Psychology of Optimal Experience*. New York: HarperCollins.

Edwardes, Dan. 2018. Move like a human: why you shouldn't exercise. *Parkour Generations*. Available online: https://parkourgenerations.com/move-like-a-human-why-you-shouldnt-exercise/ (accessed on 25 June 2019).

El-hage, Tina. 2011. Sebastien Foucan: Founder of Free Running. *The Guardian*. July 20. Available online: https://www.theguardian.com/lifeandstyle/2011/jul/20/sebastien-foucan-founder-free-running (accessed on 25 June 2019).

Fragoso, Suely, and Breno Maciel Souza Reis. 2016. Ludic Re-enchantment and the Power of Locative Games: A Case Study of the Game Ingress. In *Culture, Technology, Communication: Common World, Different Futures, Presented at the 10th IFIP WG 13.8 International Conference on Culture, Technology and Education, London, UK, June 15–17*. Edited by Jose Abdelnour-Nocera, Michele Strano, Charles Ess, Maja Van der Velden and Herbert Hrachovec. London: Springer.

Geyh, Paula. 2006. Urban Free Flow: A Poetics of Parkour. *M/C. Journal* 9. Available online: http://journal.media-culture.org.au/0607/06-geyh.php (accessed on 29 June 2019).

Hébert, Georges. 1912. *Practical Guide to Physical Education*. Translated by Pilou, and Gregg. Available online: http://stuff.maxolson.com/Practical-Guide-of-Physical-Education-1912.pdf (accessed on 29 June 2019).

Huizinga, Johann. 1949. *Homo Ludens: A study of the Play-Element in Culture*. London: Routledge & Kegan Paul.

Kidder, Jeffrey L. 2017. *Parkour and the City: Risk, Masculinity, and Meaning in a Postmodern Sport*. New Brunswick: Rutgers University Press.

Marshall, Bill. 2010. Running across the Rooves of Empire: Parkour and the Postcolonial City. *Modern & Contemporary France* 18: 157–73.

Mould, Oli. 2009. Parkour, the city, the event. *Environment and Planning D: Society and Space* 27: 738–50. [CrossRef]

Novak, Michael. 1993. The Joy of Sports. In *Religion and Sport: The Meeting of Sacred and Profane*. Edited by Charles S. Prebish. Westport: Greenwood Press, pp. 151–72.

Saville, Stephen John. 2008. Playing with fear: parkour and the mobility of emotion. *Social & Cultural Geography* 9: 891–914.

Turner, Victor, and Edith Turner. 1978. *Image and Pilgrimage in Christian Culture: Anthropological Perspectives*. New York: Columbia University Press.

religions

MDPI

Article

Redemption of 'Fallen' Hero-Athletes: Lance Armstrong, Isaiah, and Doing Good while Being Bad

Andrew R. Meyer ⓘ

Department of Health, Human Performance, and Recreation, Baylor University, Waco, TX 76798, USA;
Andrew_Meyer@Baylor.edu

Received: 15 July 2019; Accepted: 15 August 2019; Published: 19 August 2019

check for
updates

Abstract: Lance Armstrong's achievements in cycling will forever be overshadowed by his admittance of using unethical performance enhancing means to win. However, Armstrong's positive social impact of raising awareness, hundreds of millions of dollars, and support for the cancer community are undeniably noteworthy. Clearly, Armstrong's hero-savior athlete depiction in the media prior to his 'fall' was related to the social 'good' he was equally known for. This good stands in stark contrast to his demonization since. This dichotomy of Armstrong's profiling offers a unique opportunity to consider how his rise and fall reflect biblical themes of a sport celebrity. This paper explores the theme of redemption specifically presented in the book of Isaiah, as I explore Armstrong's media rendering as a fallen hero-athlete following his public acknowledgement of cheating. This manuscript provides a contextual comparison of Armstrong's story to the redemption of exiled Jews as detailed in Isaiah. Throughout the paper, I present how Armstrong has received a more profound, though less obvious or common redemption through his lifetime ban from sport. Ultimately, this article provides an analysis of a contemporary hero-athletes redemption who cycled for good, while being bad.

Keywords: Lance Armstrong; Isaiah; redemption; contemporary sport culture; exile

1. Introduction

In 2019, another fallen athlete, Tiger Woods, won the Master's Championship golf tournament. Weeks later, he received the Presidential Medal of Freedom, the highest distinction any citizen of the United States can be awarded. Through the glowing media coverage of his comeback, sporting victory, and his elevation as an American ideal, we see the ability of disgraced athletes to regain their status after a 'fall', returning to the top of their sport and in the eyes of fans. Woods' portrayal following his personal, off-the-course misdeeds over the past decade, was harsh. However, he kept playing his sport. After his latest PGA Master's win, he was heralded as an athlete who fell from the pinnacle, to the depths, only to emerge a changed and better person, worthy of praise, admiration, and national recognition. Overwhelmingly, Woods' sponsoring companies (Kennedy 2019), the world of golf (Pelletier 2019), and the general public had forgiven him; he was redeemed. But I argue through this paper that Woods has been redeemed in less a holistic and humanly impactful way, because he could remain in sport. He was able to restore his balance (redemption) as most athletes in contemporary sport are; he was able to be a popular celebrity athlete again. This kind of redemption in athletics is common. When this sport 'redemption' occurs, however, and given the kind of redemption found in the Old Testament, the person, the human that is the athlete, is left out of the equation and redemption is partial.

This article then is focused on another athlete's seeming 'unsuccessful' attempts to regain prominence in their sport, adoration of fans, sponsorship deals, and regain a general positive public opinion for themselves. This article's focus is on arguably the most vilified and despised athlete in contemporary sport and his inability to do what Woods and many others have been able to do—regain his image as a hero-athlete. Partially, this is because of the immense damage he inflicted upon himself and others within his sport over the past twenty years, and his lifetime ban from competition (Millar 2013). Lance Armstrong is now known as a cheater, liar, and fallen hero-athlete, whose collapse from global elevated status affected not only his sport, his family, and his income, but also affected those who found hope in him as a cancer survivor and his philanthropic efforts for the cancer community. Tiger Woods did bad things and hurt people around him. But millions of people held Lance Armstrong up as an idol, found hope in him, and in a very clear way 'worshiped' him. His riding and celebrity status made his cancer-related philanthropy LIVESTRONG a global success, raising hundreds of millions of dollars and awareness for cancer-related issues. In doing so, Lance Armstrong, the athlete and cancer survivor, raised something less tangible—hope—which is perhaps why so many felt harmed when he admitted to taking performance enhancing substances during his professional cycling career. As millions of people experienced this hurt, redemption was going to be difficult, if not impossible to achieve in the context of contemporary sport culture.

Armstrong's performance enhancing substance use, subsequent lies, and vilification of accusers is prolific and well documented. Our examination here is not about what he did, and the shockwaves it caused through various entities, communities, and for individuals around the world. Rather, I wish to examine the evidence of Armstrong's public story through a critical comparison of a Biblical perspective to illustrate themes of redemption in his case. This examination comes twenty years after he won his first Tour de France in 1999. There are not many who have not heard about his rise and fall.

One key issue that must be addressed at the start, is that Armstrong's transgression occurred directly related to his sport performance. The thing that made Armstrong a celebrity hero-athlete was his cycling performances and his achievements on the bike. Woods' 'immoral' actions occurred away from golf, and unrelated directly to his successful performances. Even if the stress and distractions of the negative media coverage impacted Woods' play, the improper behaviors he needed to atone for and be redeemed for were of a personal kind and not directly sport related. Armstrong's immoral behaviors related directly to that which he was famous for, namely his cycling victories. Therefore, we must be aware of the deeper connections and harm Armstrong caused due to the central connections between his performance's, his popular rendering, and his transgressions.

This paper is also not an in-depth analysis of Isaiah, but rather an attempt to bring insights of this Old Testament book to bear on a critical cultural analysis of a fallen hero-athlete in contemporary sport culture. This is not an argument for the condemnation or exoneration of Armstrong; nor is this a historiography of a book of prophets, both of which fall beyond my training and ability. The discussion presented here is part of a longer conversation I have been engaged with for the past ten years. My doctoral dissertation focused on muscular Christianity, Radical Orthodoxy, and the American cyclist who was about to come out of retirement. I first began to consider how Armstrong would be redeemed (if at all) not long after he admitted to a career built on performance enhancing substances. I followed his very public *fall* with great interest and concern. I had been a fan of Armstrong, had watched his Tour victories, and had believed in his mission for the cancer community (and beyond). Truth be told, I still am. Many friends and colleagues asked me questions about his case, his denials, and his eventual admittance. I was even asked during a tenure review when it was going to be time to stop writing about Armstrong because he was old news. I have been with this story for a decade now, a decade filled with Armstrong's triumphant return to the sport he helped globalize, watched the veil of denials slowly come down over a period of two years, the interview with Oprah Winfrey where he stoically admitted to the wrongdoing and pain he caused others. Since 2013, I have watched the punishments come down, reading about the wider anger and pain he inflicted, and erasure of his athletic achievements. What is left, what has always been at the center of it all, is a man, a human being, who at one time had fame, popularity, and success, once described as "Jesus Christ" on a bike (Pucin 2009). Now he is a person who has experienced the erasure of all that loftiness and known for his "colossal" fall (Busbee 2019). Armstrong's story can be broken into four stages (borrowing a cycling metaphor): cancer survival, professional achievement and global celebrity, the 'fall', and life since. In this simplification, the cyclist's life resembles that of the Jewish people found in the book of Isaiah: remembering the Exodus and delivery to the Promised Land, thriving and success of Israel, destruction of Jerusalem, the Temple and exile to Babylon, and eventual redemption and renewal. The 'fall' and redemption pieces of the Jewish story, is where I will focus as I relate these themes to Lance Armstrong, as a partial answer to those who keep asking me *what is Armstrong up to these days?* Using recent evidence, this examination concludes finding redemption is at hand for Lance Armstrong when Isaiah is used as an interpretive lens.

2. Contextualizing the Scope

Much could be written on each of the topics below, but in this section, I wish to narrow the scope for the reader as these concepts relate to the discussion at hand. For example, the notion of redemption could be discussed in the context of various groups related to the Armstrong case (fans, the sport, individual riders). However, as this paper is about the redemption of Armstrong as a person considering notions of redemption found in Isaiah, my review of this topic, and others, is focused toward informing this.

2.1. Isaiah

Isaiah is the book of the Old Testament that scholars argue redefines God for the Jews, and eventually Christians, from a God of creation, to a God of salvation (Anderson 1962). The God of creation is the Exodus story familiar to those Isaiah was writing for. Isaiah's messages are about the reasons for God's condemnations of the Jews in Jerusalem during their flourishing, their punishment through exile to Babylon, and the hope they should cling to through the pain and suffering of their exile (Brueggemann 1984). Isaiah writes to foreshadow punishment as a reminder to the exiled Jews to have faith, that the redemption of Jerusalem is possible, if not a guarantee (Dumbrell 1985). This is also what soteriological scholars have identified in Isaiah as shifting away from the "old" Exodus narrative familiar to Jews living in Babylonian exile to a "new" message of hope in a purified city for God's people. As Harner (1967) says "we receive the impression that the Exodus tradition provides the grounds for believing in Yahweh's new act of redemption" (p. 303). This new rendering of God is for a people who Isaiah wishes to unhinge from holding on to 'former things' or 'things of old,' namely believing God's first Exodus was final and eternal. He wishes them to focus on 'new things' that God is doing for them now, such as delivering them from exile, and His future commitments to them. Here, Isaiah is crafting an image of God as not only creator, but also as enduring savior so that all experiences of the Jewish people fit into God's eternal plan for them, and to not only cling to the original promises and past deliverances. The God found in Isaiah as savior and redeemer, not just creator, is a clear shift in the prophets work and offers a new critical perspective for the Jewish people.

A second aspect of redemption found in Isaiah that I wish to highlight is how painful the revealing of this new truth of God is and how necessary a component for redemption these are in Isaiah. Brueggemann (1984) states in his examination of Isaiah that "each of the Isaiah's articulates a specific practice of social transformation ... the text is more than a text, it is a presentation of a way through to a world of faith" (p. 91). First Isaiah is a social criticism, a judgement, "a precise exposé of cultural practice and cultural value which engage in systemic perversion" (p. 92). A new truth emerges from Isaiah's critique of ideology. "One cannot move from critique (judgement) to promise, as Childs and Clements seem to suggest, without the intervening reality of pain expressed. Second Isaiah presumed the grief and brings it to speech, even as it is evoked by the critique of ideology ... It is the suffering of exiles, the embrace of pain made possible by critique of ideology, that permits the announcement of newness" (p. 96). The pain and suffering are a necessary condition for the new relationship Isaiah sees the Jewish people having with God after their exile.

A third aspect of redemption in Isaiah that informs the focus of this paper is the hope we read throughout his book. Anderson (1962) examines the Exodus typology in Second Isaiah, again examining the difference between a God of creation and a God of salvation. If the Jews are God's chosen people, how could He allow them to suffer so greatly, through the destruction of Jerusalem, the temple, and their exile in Babylon? Isaiah's book provides his insights of God's wider plan so that hope and faith in God prevail, even if current situations conflict with previously held beliefs. Anderson (1962) states Isaiah "freed Israel's religion from the particularities of Israel's history and set forth ideas and principles whose validity is independent of the historical circumstances through which they were mediated" (p. 180). In such a rendering, any event can be examined and found to be meaningful, predetermined and part of God's larger story, a story with redemptive purpose. "According to Second Isaiah the whole course of history, from beginning to end, is set within the purpose of the eternal God, the Creator, and Sovereign" (Anderson 1962, p. 187). In doing so, Isaiah encourages his readers to remember the past, learn from it, and have *hope* in God's power to use any situation as evidence of His redemptive ability. "The Exodus is a guarantee that Yhwh will redeem his people, for that event demonstrates that he has the wisdom and power to accomplish what he purposes. Second Isaiah spoke to a people in exile, in despair about the meaningfulness of their history and about Yhwh's power to give them a future. The prophet's intention is to awaken their confidence by proclaiming that Yhwh is the only Lord of history, for he accomplishes what he announces. Israel's redemption will surely come ... (55: 10–11)" (p. 189). This new exodus found in Isaiah "will be a radically new event. It will surpass the old exodus not only in wonder but also in soteriological meaning, as evidenced by the theme of divine forgiveness, which runs through the whole of his prophecy, or by extension of salvation to include all nations" (p. 191). Through the despair caused by their destruction and exile, Isaiah constantly provides the alternate, assuring his readers of the hope of God's eternal presence and favor, in the darkest of times, and in the unlikeliest of events.

What is important about Isaiah's themes of redemption here, is that firstly he is redefining the path of God for the Jewish people, as one from creation to that of salvation. In this reinterpretation, he is critiquing and challenging the way the Jewish people understand God and their relationship with Him. This social critique and new understanding, according to Brueggemann (1984), is essential for the revelation of truth, and a moving away from the false truth of "positivistic claims" by those with power (pp. 93–94). Secondly, inherent to this critique is that any new understandings of the truth will be painful. While the initial challenge to the dominant version of truth will cause the suffering of individuals and groups, Isiah reveals that this will be necessary for the redemption of the Jewish people. Finally, through this painful critique of old ways, Isaiah continually offers visions of how the "new" way leads to redemption and a life filled with hope. We will see these three themes later in the paper as they relate to Lance Armstrong and his path towards redemption as a person, and not just a hero-athlete.

2.2. A Moral Role of Sport and Hero-Athletes

Sport and religion have historically shared common elements of human communal needs and a morality. Guttmann (2004) focuses on ritual, for instance, suggesting the sport activities in pre-modern cultures "were often—perhaps usually—embedded within or aspects of religious ritual" (p. 7). Other scholars have focused on sport and religions' participation with nature (Sansone 1988) or formation of community bonds (King 2006). Shirl Hoffman (1992) writes in his work that "sport has long been regarded as a shaper and reinforcer of values deemed critical to the maintenance of American society," (p. 6), and display ritualistic "properties," like religious rituals, that "reinforce the community's commitment to society's core values" (p. 7). Some scholars suggest that sport is a "type" of religion, for example, Novak (1992) who claims sport "is not a religion in the same way Methodism, Presbyterianism, or Catholicism is a religion," but that sports generate a form of civil religion, as these formal religions "are not the only kinds of religion" (p. 35). Mathisen (1992) argues that if sport is a civil religion, then the core of sport is a folk religion because civil religions have the tendency to come and go with passing popularity (p. 18). Using Linder's (1975, p. 401) definition, Mathisen says "folk religion is a combination of shared moral principles and behavioral customs," and these principles and customs "emphasize the common religion of a people as it emerges out of the life of 'the folk'" (Mathisen 1992, p. 19). Members of a folk religion "affirm their beliefs and practice their rituals" making up a "collective conscience" (p. 23). Albanese (1981) suggests even a third view of sport as a "cultural" religion. Such "religious forms include sacred stories, rituals, moral codes, and communities … " (p. xxi). I contend that in contemporary sport cultural, sport takes on and fulfills all the various roles that Novak, Mathisen, and Albanese discuss. Sport thus becomes a religious and spiritual experience, with moral and ethical values, and includes hero-role models who display the ethos of the system and perpetuate an ongoing dialogue (within its confines and for the broader culture) of what it means to be good, right, and true.

As a result of the decrease in traditional religious adherents, many find the spiritual and religious messages of sport and athletes powerful and attractive (Nelson 2009). In their understanding of the nature of postmodern theology, Radical Orthodoxy scholars, such as Milbank et al. (1999), support notions that alternate areas of culture operate in spiritually and religiously significant ways. As such, sport has become a self-sustaining entity with diverse media outlets, becoming an organic whole that disseminates moral value. Through this organic and ubiquitous contemporary role, sport has become a site of religious and spiritual meaning, where athletic hero/icons reflect the hegemonic moral and ethical values deemed important today. Understood in these terms, contemporary sport is not a *type* of religion, but rather a cultural activity of religious and spiritual meaning in the lives of many. As Sydnor (2003) suggests "our studies and conclusions … we might boldly answer that the developed world's obsession/fascination with … sport-related productions and representations is the result of individual and societal emptiness that is only fulfilled by God" (pp. 26–27).

For Oriard (1982), however, a sport hero is largely a "prowess hero" rather than an "ethical hero" (p. 30). This tracks with Allison and Goethals's (2011) conclusion that "[a]lthough heroism can be used in either conscience or competence alone, most of the examples … combine both qualities" (p. 200). Any figure or activity that dominates our attention and lives, as sport and popular athletes do today, has the trappings of idolism. As such, the heroes and icons popular in sport are empty vessels of true meaning in our lives. Whatever the justifications, often, we do hold these individuals up as moral ideals and standards bearers, forgetting that our current cultural practices and mores put athletes on pedestals, demand they be more than human in their behaviors, and thus we are implicit when they 'fall'. Meyer's (2012) work on muscular Christian reflections of Lance Armstrong in the media highlights evidence of contemporary hero-athlete worship. But as Williams (2009) states, there is an "absurdity of athlete worship" due to "the mere fact of being well known is not enough to transform athletes into moral standard bearers" (p. 13). As is often the case in contemporary culture, media is deeply involved in encouraging the sport system to rid itself of fallen sport heroes, as well as exuberant and celebratory when they return to prominence, promoting the age-old story of beating the odds and depict them as 'worthy' icons. Little scrutiny is ever given to the culture of sport itself, its socialization toward deviant behavior, and the media's role in promoting these themes.

2.3. Redemption

Redemption is the righting of wrongs, errors, or sins, and coming back in line with the previous or established operation of a system (religious or otherwise). Simply defined, redemption is "the process of righting a wrong and restoring balance" (Scholes and Sassower 2014, p. 131). Crimes against society are met with penalties which, upon fulfillment, individuals are said to have been redeemed. There are of course biblical perspectives of redemption. In the Old Testament, for the Israelites, redemption came when there was "recognition of the nature and gravity of" the "mistake and acknowledgment of responsibility for their sins" (Scholes and Sassower 2014, p. 132). Thus, two key themes are necessary in the context of Jewish redemption: acknowledgment a wrong has been committed and taking personal responsibility for that sin.

For professional athletes, depending on their mistake, redemption can come in many forms (ejection, suspension, fines, etc.). However, all redemptive efforts must involve an acknowledgement of wrongdoing and an "explanation of why the mistake was made" (Scholes and Sassower 2014, p. 137).

Regardless of the sin, harm, or infraction, redemption is granted by "the one sinned against," (i.e., God or others) "though neither is obligated to do so" (p. 133). In contemporary sport culture, redemption is difficult because rarely is there a specific person an athlete can apologize to and garner redemption. For example, American football quarterback Michael Vick went to prison for engaging in an illegal dog fighting ring. After serving his time in prison and paying fines, Vick returned to his career in the National Football League (NFL). In the eyes of the NFL, and in the eyes of the state, he had been redeemed. But the throngs of fans may never have given him that same "return to balance" in their perceived relationship with him. In fact, this type of redemption would be impossible. It is clear then that redemption for contemporary athletes, especially high-profile celebrity athletes, is not feasible.

As this paper is focused on comparing the redemption described in Isaiah with the story of Lance Armstrong, I am not as concerned with clarifying who and what of Armstrong's redemption. In Isaiah, we read how after their exile, the Jews were in fact not returned to the same Jerusalem, as it had been destroyed. The redemption following their exile was not a direct restoration to balance with Jerusalem (their original place) and with God, but rather to a new place and a new relationship with God. Additionally, even though all Jewish people were rescued from exile, not all 'redeemed' entered the promise of the 'new' Jerusalem or a restored relationship and balance with God. Those who returned to the city but did not have faith in God are described in Isaiah as the "wicked," and those who returned from exile with hope and a revived faith in God were called "servants" (Roberts 1982, p. 136). Examined below are the similarities of Armstrong's walk towards redemption with that of Isaiah's description of the redemption of the Jews from Babylonian exile, and how the lifetime barring from professional sport assisted in his holistic and more meaningful redemption as a person.

3. Comparative Analysis: Isaiah and Armstrong

The book of Isaiah is a conglomerate and many scholars have noted the complex nature by which the book became cannon, claiming with a high degree of assuredness that the Prophet Isaiah of Jerusalem most certainly did not compose the entire book (Clements 1982, p. 127; Dumbrell 1985; Anderson 1962; Roberts 1982). This paper is not a soteriological analysis of Isaiah but rather an examination of the meaningful connections between the warnings, failings, destruction, exile, and restorative promises found in Isaiah and the 'fall' and redemption of Lance Armstrong. This section provides a comparative analysis of Isaiah, in general, to Lance Armstrong's story of 'falling' from the loftiness of sport celebrity and his case for redemption. My focus will be on presenting an overview of the sections of Isaiah I wish to relate to Armstrong and then do so with evidence of his journey.

It is helpful to understand how some scholars have broken the book of Isaiah into groupings, to help readers clarify different sections and overall themes of the book. Brueggemann (1984) reads Isaiah in three parts (as do most scholars); chp. 1–39 as "*a critique of ideology*," chp. 40–55 as "*a public embrace of pain which leads to hope*," and chp. 56–66 as a "*release of social imagination*" (italics in original) (p. 102). Dumbrell (1985) reads the book with "an overmastering theme . . . of Yahweh's interest in and devotion to the city of Jerusalem" (p. 112). He also breaks the book into parts, with chp. 1–39 representing a picture of Jerusalem in decay, "whose sacrifices cannot any longer be accepted and whose prayers must be turned aside" (p. 112). He reads the second half of the book (chp. 40–55 and 56–66) as describing the gradual change which occurs following the exile to Babylon—punishment for the sins detailed in the first half of the book. Other authors have explored specific themes found in Isaiah, and especially chp. 40–55, focused on turning from creation to salvation, are most apparent (Anderson 1962; Harner 1967).

3.1. Arriving at and Thriving in a Promised Land—Moving towards the 'Fall'

Clements (1982) identifies that the first chapters of Isaiah provide examples of a city that was flourishing, but with a population that had turned away from God. The prophet's warnings reveal Jerusalem's ignorance as his prophecy is described as falling on "deafness and blindness" (Clements 1982). Isaiah highlights how the kings of Judea believe that if they turned to regional allies, such as Egypt and Babylon, rather than God, they could defeat the Assyrians, an immediate threat in Isaiah's book. Isaiah highlights Jerusalem as a thriving city, and in ways that very much demonstrated prior promises God had made to the Jewish people. He had delivered them from Egypt and provided Judea as the Promised Land. He bestowed the blessings of military victory, wealth, and security from enemies. By the time most scholars believe the prophet Isaiah of Jerusalem was writing, beginning in about 737 or 736 B.C.E. (Clements 1982, p. 120), the people of Jerusalem were thriving. Isaiah, however, saw dangerous trends coming, which culminated in his prophecies. He warned the continuation of Jerusalem's prosperity is "Yahweh's guidance … not the deft political kingship or conjunction with foreign alliances … " (Dumbrell 1985, p. 115). The first chapters of Isaiah are "a thoroughgoing indictment of the failure of Israel to be the people of God and a rejection of Jerusalem," and provide "a threat directed to a Jerusalem society given over to pride and idolatry" (p. 113).

Upon this overview of Isaiah, already some comparisons with Lance Armstrong's story can be made. Armstrong had survived Stage 3 testicular cancer, and had been regranted his health, life, and cycling career. His story of survival can be read like that of the Jewish Exodus from Egypt, who were delivered from slavery (death) to the Promised Land (life). Once Armstrong returned to cycling, he lived a profoundly successful and celebratory life; he lived in the 'promised land' of sport success. He was victorious on the bike, and popular as an athlete, celebrity, and philanthropist. He attributed his successful survival of cancer to the medical doctors and treatments, citing science as the reason he survived. A reliance on science, medical treatments, and drugs continued throughout his cycling and philanthropy. Like the Jews of Isaiah's time, Armstrong moved away from an internal need for others, a deep sense of self, and creating meaningful relationships toward specific calculated and rational alliances that kept his hero-athlete narrative intact. This also is comparable to Isaiah as the prophet was aware of the danger of false narratives promoted by powerful people and "Isaiah does a discerning critique of the power of propaganda" (Brueggemann 1984, p. 94). Much like the Jews in Jerusalem, Armstrong made strategic alliances with others that broke the moral code of his sporting world in order to falsely maintain his elevated status as hero-athlete.

Starting with his 1999 Tour de France victory, Armstrong was plagued by suspicion and claims of illegal training substances and methods by a small few, in and out of sport media. He of course vehemently denied these, as is well documented, attacking, suing, and demonizing those who questioned the purity of his performances and achievements. His followers, supporters, teammates, coaches, and others in the media defended, promoted, and protected him from the truth of these allegations. The small number of detractors were like Isaiah in Jerusalem who called for others to awake to the reality of their actions and behaviors. Like those of Jerusalem, the warnings and proof of Armstrong's illegal performances fell on deaf ears and blind eyes. Lance Armstrong rode to the most ever Tour de France wins between 1999 and 2005 (seven consecutive) and retired from the sport he had helped globalize to focus on his philanthropic efforts with the cancer community. Comparatively, when Jerusalem was under assault from Assyria, the people of Jerusalem prayed to God to save them, which He did. They felt a sense that God was with them and they could continue in the ways Isaiah was critiquing. Armstrong also seemed to have beaten all the allegations and speculation, and could go on with life, never having been caught cheating.

As with all tragic stories, however, Armstrong did not leave the sport for good. In 2009, he decided to come out of retirement and ride in that year's Tour de France. Armstrong has identified in several interviews that if he had not come out of retirement, he would not have been caught (Levy 2012). After riding in 2009 and 2010 and retiring once again from the sport, by 2012 a resurgence of claims against him had resurfaced. Comparing once again Isaiah, after the Jews had survived the Assyrians, and with their kingdom reduced in size, the Jews continued around 100 more years until eventually the prophet's forecast came to be realized (well after his death). Babylon razed the city of Jerusalem, destroying the sacred temple, and sent the Jewish people into exile. In both examples, each resided in a 'promised land,' but were not living properly. Both engaged in improper behaviors, contrary to the expected practices of their social context, wrapped up in their self-reliance, deafness, and blindness to truth by silencing those who dissented against them (Brueggemann 1984, p. 94). This eventually led to their destruction and exile.

3.2. The Suffering: Destruction and Exile

The exile of the Jewish people is foretold as the result of their sinful actions and loss of faith in God in the first half of Isaiah (chp. 1–39), with further justification and rationale provided in the second half (Harner 1967, p. 298). God levied this punishment so that His people and their wicked and sinful ways could be forgiven (Roberts 1982, p. 136). Chp. 40–55 are focused on the return of the exiles from Babylon and the restoring of their new relationship with God through the rebuilding of a new Jerusalem and temple. What we see in these later chapters is the hope realized following the painful suffering of a people who had lost their way.

If we relate the exiled experiences of the Jewish people with Armstrong's 'fall' from hero-athlete, we can again come to appreciate how the painful suffering he experienced compares with that described in Isaiah. While it is not my contention that he immediately came to be redeemed when he lost his sponsors, fans, and ultimately his affiliation with LIVESTRONG, there is ample evidence that his punishments and exile caused him to truly suffer over an extended period, a necessary experience for redemption.

After a lengthy investigation by the United States Department of Justice in February 2012, which produced no criminal charges being filed, it looked as if Armstrong was exonerated. Yet, in June of that same year, his case was picked up by United States Anti-Doping Agency (USADA). As before, Armstrong denied the allegations and sought to fight against these new cases. However, by August of 2012, with many former teammates testifying and granted plea deals, he announced he would stop fighting the charges brought by the USADA. He was handed a lifetime ban on all competitions and the disqualification of all his competitive results from August 1, 1998 onwards (USADA 2012). By October, Armstrong could no longer hide from the truth as the USADA made public a 202-page document of evidence. On 17 October, he stepped down as LIVESTRONG chairman and Nike, 24-h Fitness, Easton-Bell (Giro helmets), Trek bicycles, Honey Stinger, Anheuser Bush, Oakley Sunglasses, and Radio Shack terminated their sponsorship contracts with him (Rishe 2012; Rotunno 2012). On 22 October, the International Cycling Union (UCI) upheld the USADA's decision to strip him of his competitive record, including his seven Tour de France wins, and on 12 November Armstrong stepped down from the board of LIVESTRONG—ending any formal relationship with the foundation he created fifteen years earlier. Finally, on 17 January 2013, the first of a two-day televised interview was broadcast in which Armstrong admitted to Oprah Winfrey the truth of the charges against him for the first time and attempted to justify his actions.

Significant in his redemptive suffering, Armstrong's lifetime competition ban means that he could not compete in any sanctioned events, in any sport. Over the next few years, Armstrong came to realize the extent to which he could not compete in his new state of exile. In April 2013, he attempted to swim in a local United States Master's Swimming (USMS) meet in Austin Texas. He registered believing that because the meet had no professional outcomes, nor prize money, that his involvement was allowed. But when the World Anti-Doping Agency (WADA) heard of his registration, they contacted the International Swimming Federation (FINA) who asked the local sponsors of the meet to bar his involvement. Armstrong said "I was told all along that I was more than welcome to compete in masters meets by U.S.M.S. Then all of a sudden, I'm not welcome? I don't get it." (Macur 2013). He was also stopped a year and a half later from riding in his former teammates George Hincapie's *Gran Fondo* ride. While Armstrong was not involved in any competition, the fact that results from the event were submitted to the National Ranking System meant that Armstrong was not allowed to participate (Farrand 2014).

As Armstrong was clearly exiled from sport and competition, he also suffered an exile from those who had found meaning in his hero-athlete status. Along with his exit from LIVESTRONG, there was well documented backlash against him as a person and inspirational figure. Internet videos were posted of people cutting their yellow LIVESTRONG bracelets (Rapoport 2013), burning his books, and many attempted to return LIVESTRONG items to stores for refunds (Kelly 2013). Any review of comments posted to Armstrong's social media at the time, and since, reveal the fervent anger and pain his revelations caused.

Since his 'fall,' Armstrong has existed in relative exile, facing the fallout of his actions and living a life away from the sport and spotlight that made him a hero-athlete. The one thing that Armstrong was not allowed to do was return to professional sports. I contend this has been redemptive in that he was not allowed to return to the activities that placed him in exile in the first place. Here, we can apply Isaiah once again, as the Jewish exiles did not return to their former city, which had been razed by the Babylonians. Rather what Isaiah describes is the return to a 'new' Jerusalem, where the lessons learned from their exile restore them to God's favor and gives them hope for a better future.

3.3. Beginnings of a New Hope

By the end of Isaiah (chp. 56–66), details are provided about the redemptive hope described earlier in the book. We read about the promised reward of the faithful, the "righteous" servants and how they will be separated from those who continue to repeat the sins and errors of the previous generations, "the wicked" (Roberts 136). Those who see the new Jerusalem for what it is, another promised fulfilled by God, will live in hope. It is clear, however, that the new hope is not extended to all who returned from Babylon. Isaiah states that God will extend this newness and hope only to the faithful, not the whole nation (Dumbrell 1985, pp. 126–27). Here, the limits of redemption in Isiah where God redeems everyone, but not all will see or live the life of redemption, is apparent. Hanson (1975) explains how not all Jews will come to receive the benefits and redemption of their return from exile. Dumbrell states that God has divided the exiles into "'my people'" and "'those who forget my holy mountain'" (65:11). What is illustrated in the exiles' return from Babylon is that in this 'new' Jerusalem and relationship with God, some receive full redemption while others, though returned from exile, do not. The servants will profess their sins and lament the evil they see, asking for God's forgiveness, while the wicked returned from exile and believe themselves to be redeemed, returning to 'old' ways.

As I have said earlier, Armstrong is different from other fallen athletes in that he was never allowed to compete again and regain his former hero-athlete image through competition. I contend that this renders his portrayal to that of a servant. Other forgiven athletes are 'redeemed' because they go back to sports' elevated status and are reflective of the wicked described in Isaiah. They are wicked because they return to their previous elevated status as athlete, and not necessarily because they re-engage in improper behaviors. They return to a system that is flawed, whereas Armstrong was never allowed to re-enter the flawed system. This revelation has slowly come to be observable as I have documented Armstrong's reactions and interactions in popular culture more recently.

Many of Armstrong's first reactions were less than humble or conciliatory. During the early part of his exile, as noted above, he attempted to participate in events he was barred from, claiming he was being singled out in a sport riddled with doping scandals. "Despite moving away from professional cycling Armstrong remains convinced that he was made some kind of scapegoat and has paid for the sins of his generation and even the generation that proceeded his career ... " (N.A. 2014). Armstrong consistently presented an air of righteous contempt for the way he was banned and exiled. In interviews, he expresses bitterness, anger, and defiance in the face of his 'fall.' When offered opportunities to help investigators, he was less than willing at first. He called authority figures at USADA names, labeling them in unflattering ways. Many said of his initial apology with Oprah that he never cried or showed emotion, therefore he could not be sorry for what he did, only that he was caught (Karlinsky and Castellano 2013).

It is difficult to tell what is going on in someone's heart, or what their true feelings are, from an interview. I suspend my own judgement of Armstrong's early public presentations of himself through the media as he struggled with his *persona non grata* reality. But his defiance and the chip on his shoulder is well documented in articles and interviews. Sanderson and Hambrick (2016) write of

Armstrong's attempts at apologizing, asking whether his was a case of an authentic apology. Did he really regret what he did, or were his attempts at apology attempts to repair his tarnished image? Did he really feel bad about the pain he caused, or just that he was finally caught? I contend we can never say for sure. However, analyzing the evolution of his apologies and the regret he displays, one can see how his regret today is different from the regret he espoused in the early years of his exile.

After two years of exile, opportunities for redemption of the contemporary sport culture ilk, especially related to cycling's governing body, started to appear. In late 2014, then president of the International Cycling Union (UCI) Brian Cookson stated "I think there is potential for redemption for him and anyone, really. I think it all depends on what (Armstrong) said to the commission and if he was prepared to talk about his or other people's involvement and whether he's genuinely contrite and deserving of redemption" (Windsor 2014). Armstrong stated that he wished to work with the Cycling Independent Reform Commission (CIRC), a commission formed to investigate cycling's doping past. In doing so, the commission had the authority to offer reduced sentences to those who fully cooperate and testify. Armstrong stated, "I've always said I will make myself fully available to an international commission tasked with helping our sport heal and move forward after multiple generations of rampant doping" (Benson 2014).

In 2016, after serving four years of his lifetime ban, and under very limited press release, Armstrong was granted an allowance in which "he can compete in a sanctioned event at a national or regional level in a sport other than cycling that does not qualify him . . . to compete in a national championship or international event" (Schrotenboer 2016). Armstrong has also described his attempts at apology and seeking forgiveness and redemption by those he individually hurt. "I've apologised multiple times [to Betsy Andreu] . . . What I've learned is you can't force someone to accept an apology . . . I've traveled the world to make it right with these people . . . Not only did I say [sorry] . . . but I meant it. I don't know what else I need to do" (N.A. 2016). In 2018, Armstrong settled for $5 million with the U.S. government for defrauding taxpayers and lying to the government. (Schrotenboer 2018).

I argue, however, that these examples of lifting certain aspects of a lifetime ban or paying the government, are not examples of the kind of redemption reflected in Isaiah. While Armstrong has made attempts to apologize on national television and to individual persons, his circumstances fall to the limitations detailed earlier, that an athlete can never speak to everyone and ask for forgiveness, to attain redemption. By the contemporary sport culture standards, Armstrong may have gotten his 'license to ride' back and settled with the federal government, but in a May 2019 interview (NBCSports.com), we can see a deeper redemption in the answers Armstrong provides, demonstrating that he has truly suffered and has come to a more humble and profound understanding of himself and his predicament, reflecting Isaiah's description of the humble servant after exile.

3.4. Glimpses of Redemption

In May of 2019, an interview was published in which Armstrong answered questions of where he was now (Brennan 2019). Many familiar questions were posed, but the interesting thing about this interview was the depth to which the usually stoic Armstrong went with his answers, and his emotional vulnerability. Three times in the interview (a familiar count for Biblical fans), Armstrong answered the question if he could do it all over, what would he change. His answer was "nothing"; he would not change a thing (Gaydos 2019; Busbee 2019). In previous interviews, he had stated that people did not like this answer. But in the 2019 interview, he said he would do it all over again, because it led him to sitting in that chair, having experienced all the pain and suffering of his exile. He was unburdened, he was free, he had learned his lessons from his mistakes and he no longer looked to others to justify his actions. He stated that if he had not done what he did, experienced it all, he would not have come to be a better person (Busbee 2019). Throughout the interview, we see a man who has been redeemed. Not by the millions of fans, the UCI, nor the U.S. government. Lance Armstrong was redeemed by a new understanding of the pain and suffering he experienced and has moved into a place where he is experiencing the newness of a life of hope. Lance Armstrong reflects the version of redeemed exiles in Isaiah that are servants of God, and not those granted a return from exile but return to a sport culture that inherently encourages and develops 'wicked ways', even if they never repeat bad behaviors. By not returning to sport, Lance Armstrong reflects Isaiah's image of a redeemed servant through the pain, suffering and exile he experienced.

4. Three Themes of Redemption

There are three points with which I wish to conclude this comparison; they are related to the role of contemporary sport culture, redemption, and Biblical themes from the book of Isaiah found in the popular media representation of Lance Armstrong. These themes echo Brueggemann's (1984) analysis of Isaiah, which I contend help sport scholars understand the global challenge of forgiving sport celebrities when they err in and out of sport performances.

The first point I wish to make here, as alluded to earlier in the paper, is that sport organizations, sport media, and the general public must rethink what it means to redeem athletes. As I have stated throughout, the common practice today is for athletes who make mistakes, such as Armstrong, to serve a certain punishment after which they can play again. When an athlete is found to have acted or behaved in a manner contrary to the law, social norms, or moral codes, we often think of them as redeemed when they return to and succeed in their sport (Scholes and Sassower 2014). However, if we are to follow the path toward redemption laid out in Isaiah, what we must consider is if returning athletes to sporting environments, that perhaps encouraged their behavior in the first place, is truly redemptive. For Armstrong, the lifetime ban 'saved' him from his return to professional sport and thus freed him from the lie he was living. It also forced him to look elsewhere for meaning in his life, in his case therapy and self-evaluation (Brennan 2019). But the fame, accolades, and elite status returned to Tiger Woods following his 2019 Master's victory, or Michael Vick, could very well lead to similar ideations of elevated status that caused transgressions in the first place. I am not claiming that any time an athlete gets in trouble, they should be banned for life. I am asking the reader to admit that we celebrate 'fallen' athletes when they return, because we watched them 'fall' and then had the gumption to return and perform. To me, this seems like Isaiah's descriptions of the wicked who return from exile, only to engage in the same behaviors and under similar circumstances as they did before. They misbehave, are punished, and then return to play again. Whether or not they re-engage in similar unethical behaviors, professional athletes exist in a sport culture that nevertheless provides opportunities for, even perhaps encourages, unethical behavior. We must, as Brueggemann contends, follow the calls of Isaiah for social criticism of the way 'fallen' sport stars return to play and examine the contemporary culture of sport before we make any strong claims that athletes who return to play have been redeemed. Armstrong was not allowed to do that, and I contend he is better as a person because of it.

Secondly, consumers and scholars of sport should acknowledge the pain and suffering as essentially necessary for redemption. When an athlete pays a fine, or serves a suspension, we often have not inflicted adequate pain and suffering to achieve the desired effect of true redemption. When an athlete returns to their sport following an infraction, and regardless of the degree of their punishment, an opportunity to return to their 'old ways' is ever-present. Lance Armstrong experienced what many deem to be adequate pain, as his punishment reflects an ultimate end that allowed his redemption away from the context of sport. In his 2019 interview with Mike Tirico (Brennan 2019), he states that had he not doped, lied, and been such a hurtful person he would not have been so punished, and thus he would not have learned the lessons he has through the process. Because he was not allowed to compete, he never could be redeemed through sport. This was true pain as it effected his career, finances, social status, and individual sense of self. His full redemption would not have been possible if he had been allowed back to sport. Without the painful suffering, he would not have learned the deeper lessons and become a better person. In relation to Isaiah's humble servant, we can also understand that through his story, Armstrong is an example of painful suffering leading to redemption. Not redemption in sport, but redemption as a person, a human who experienced deep pain and learned a new truth about what he did and how he has been changed by that truth. As we learn in Isaiah, God uses pain and suffering, through the exile of His people, to show them a new and better way to live. In Isaiah, we read "commentary explaining how this return has been achieved, namely though the ministry of the servant who has suffered so extremely, 52:13–53:12." (Dumbrell 1985, p. 126). Armstrong needed to suffer so that a new truth could be revealed to him and others. Helping an investigation, being contrite, paying a fine, or sitting out a few games are how other athletes have been 'redeemed' by a deeply flawed contemporary sport culture. Athletes only see the path to return to their elevated status at the sport table, and the culture of contemporary sport reinforces this. Again, I contend Armstrong's lifetime ban saved his chances for holistic redemption of a different, soulful, and more meaningful kind. It allowed him to suffer so greatly that he was redeemed as a person, not just as a hero-athlete.

This critique of contemporary sport culture's false redemption and the necessary painful suffering of individual athletes to achieve Biblical redemption, point to a third redemption theme for Armstrong; there is *hope* for those in despair. Armstrong is an example of a 'fallen' athlete, who was denied any chances of redeeming himself through sport and has shown himself to be better for it. He symbolizes the hope that there are new ways of imagining redemption (Brueggemann 1984, p. 102) for athletes after a great 'fall.' Paying a fine and returning to the pitch is all well and good. But the full extent of Armstrong's suffering demonstrates that there can be even greater redemption for an athlete who can never return to their sport. Unfortunately, this hope is lost in contemporary sport culture, where athletes, the media, and fans see a ban, punishment, or fine as another obstacle to overcome, and never truly learn lessons found in the deep pain and suffering that leads to the kind of redemption found in Isaiah's humble servant. These humble servants suffered through the exile and came back to live a life of newness and hope. This is observable in the case of Armstrong, who has articulated this change and accepted that he would not change a thing, even though it was hard and painful. It is precisely because it was hard and painful that he has emerged new and filled with hope. His example gives hope that there is redemption beyond the field of sport for fallen athletes, and it is a hope that sport scholars and those in sport media should celebrate and look for when the next athlete or sportsperson returns to their previous sport life and is said to have found redemption (Wilson 2019).

Armstrong could not be forgiven nor redeemed in the first years of his admittance; he had not begun his exile and painful journey of punishment. But after the lasting pain, from sources he could not anticipate, we see his redemption as a person and as described in Isaiah. For as Brueggemann (1984) states, "the hope is both permitted and required only by the suffering faced and claimed by the exiles ... And where there is no guilt and grief, there will not be comfort spoken ... The hopeful poetry of Second Isaiah is scarred and is spoken by one who knows" (p. 96).

5. Conclusions

Lance Armstrong represents a complicated figure in the mythos of contemporary sport. His story is filled with evidence of a biblical redemption found in the book of Isaiah. The deep pain and hurt experienced in both cases also lead one to appreciate and conceive of the truth that emerges. Stemming from the discussions presented here, future scholarship should focus on critiquing the nature of professional sports, and particularly the sport of cycling, that encourages athletes towards unethical means of achieving desired performances. It is also helpful to remember at the end here, that God did not return His chosen people from exile just for themselves, but for a "deeper soteriological meaning, and with world-wide implications" (Anderson 1962, p. 194). This truth is that only through deep pain and suffering, after a transgression has taken place (whatever the context), can one come to be redeemed as a person. Armstrong's May 2019 interview reveals his acceptance that "suffering need not be shunned but can be received as a way to live that opens the future" (Brueggemann 1984, p. 97).

Through Armstrong's suffering and exile, we observe his evolution for explaining his actions; from defiant, to angry, to contrite, to smug, to nervous, to calm and genuine. He is right when he acknowledges that people do not like it when he says he would do it all over again. But given the context of Isaiah, his evolution reveals the truth that without pain and suffering through an exile, he could not learn, grow, and find redemption. Contemporary sport culture tells us that redemption for athletes culminates when a 'fallen' athletes returns to play and is assured when they win. But deeper, more meaningful, redemption of the person requires suffering the kind Armstrong experienced. He had to fall and suffer so that he could share his story and experience to show us all how to hope for a better future for sports, athletes, and life. Brueggemann (1984) suggests "it is the very act of exile, lamentation, guilt and grief which now is overcome by the act of embrace" (p. 99). Only in pain and exile can one imagine a new reality and hope for a brighter future. We should all see Lance Armstrong as a person worthy of redemption and as a reminder of the good that can always be found from the bad.

Funding: This research received no external funding.

Conflicts of Interest: The author declares no conflict of interest.

References

Albanese, Catherine L. 1981. *America: Religions and Religion*. Belmont: Wadsworth Publishing Company.

Allison, Scott T., and George R. Goethals. 2011. *Heroes: What They Do and Why We Need Them*. New York: Oxford University Press.

Anderson, Bernard W. 1962. Exodus Typology in Second Isaiah. In *Israel's Prophetic Heritage: Essays in Honor of James Muilenburg*. Edited by Bernard W. Anderson and Walter Harrelson. Manhattan: Harper & Brothers, pp. 177–95.

Benson, Daniel. 2014. Lance Armstrong Set for Second Meeting with CIRC. *Cycling News*. October 28. Available online: http://www.cyclingnews.com/news/lance-armstrong-set-for-second-meeting-with-circ/ (accessed on 28 October 2014).

Brennan, Daniel. 2019. WATCH: Lance Armstrong's full Interview with Mike Tirico. *NBCSports.com*. May 29. Available online: https://sports.nbcsports.com/2019/05/29/watch-lance-armstrong-full-interview-with-mike-tirico-video-tour-de-france-cycling/ (accessed on 7 June 2019).

Brueggemann, Walter. 1984. Unity and Dynamic in the Isaiah Tradition. *Journal for the Study of the Old Testament* 29: 89–107. [CrossRef]

Busbee, Jay. 2019. Lance Armstrong in NBCSN interview: 'I wouldn't change the way I acted'. *Yahoo! Sports*. May 30. Available online: https://sports.yahoo.com/lance-armstrong-in-nbcsn-interview-i-wouldnt-change-the-way-i-acted-150314078.htm (accessed on 7 June 2019).

Clements, Ronald E. 1982. The Unity of the Book of Isaiah. *Interpretation: A Journal of Bible and Theology* 36: 117–29. [CrossRef]

Dumbrell, William. 1985. The Purpose of the Book of Isaiah. *Tyndale Bulletin* 36: 111–28.

Farrand, Stephen. 2014. USA Cycling Stops Armstrong Riding the Gran Fondo Hincapie. *Cyclingnews*. October 23. Available online: http://www.cyclingnews.com/news/usa-cycling-stops-armstrong-riding-the-gran-fondo-hincapie/ (accessed on 23 April 2014).

Gaydos, Ryan. 2019. Lance Armstrong Says He 'Wouldn't Change a Thing' about Doping Scandal that Cost Him 7 Tour de France Titles. *Fox News*. May 24. Available online: https://www.foxnews.com/sports/lance-armstrong-no-regrets-doping-tour-de-france (accessed on 7 June 2019).

Guttmann, Allen. 2004. *Sport: The First Five Millennia*. Boston: University of Massachusetts Press.

Hanson, Paul. 1975. *The Dawn of Apocalyptic*. Philadelphia: Fortress Press.

Harner, Ph. B. 1967. Creation Faith in Deutero-Isaiah. *Vetus Testamentum* 17: 298–306. [CrossRef]

Hoffman, Shirl J. 1992. *Sport and Religion*. Champaign: Human Kinetics.

Karlinsky, Neal, and Anthony Castellano. 2013. Lance Armstrong May Have Lied to Winfrey: Investigators. *ABC News*. January 18. Available online: https://abcnews.go.com/US/lance-armstrong-lied-oprah-cover-crimes-investigators/story?id=18245484 (accessed on 20 January 2013).

Kelly, Jon. 2013. Should Buyers of Lance Armstrong's Books Get a Refund? *BBC*. February 4. Available online: https://www.bbc.com/news/magazine-21250032 (accessed on 27 March 2013).

Kennedy, Merrit. 2019. Tiger Woods Rises Again—And Sponsors Are Celebrating His Resilience. *NPR*. April 15. Available online: https://www.npr.org/2019/04/15/713443562/tiger-woods-rises-again-and-sponsors-are-celebrating-his-resilience (accessed on 20 April 2019).

King, Samantha. 2006. *Pink Ribbons, Inc.*. Minneapolis: University of Minnesota Press.

Levy, Dan. 2012. Lance Armstrong Let Pride and Hubris, Not Doping, Ruin His Legacy. *Bleacher Report*. August 24. Available online: https://bleacherreport.com/articles/1309680-lance-armstrong-let-pride-and-hubris-not-doping-ruin-his-legacy (accessed on 20 September 2012).

Linder, Robert D. 1975. Civil Religion in Historical Perspective. *Journal of Church and State* 17: 399–421. [CrossRef]

Macur, Juliet. 2013. Disappointed Armstrong Stopped from Competing in Swimming. *New York Times*. April 4. Available online: https://www.nytimes.com/2013/04/05/sports/cycling/armstrong-plans-to-enter-swimming-competition.html (accessed on 20 April 2013).

Mathisen, James A. 1992. From Civil Religion to Folk Religion: The Case of American Sport. In *Sport and Religion*. Edited by Shirl Hoffman. Champaign: Human Kinetics.

Meyer, Andrew R. 2012. Muscular Christian themes in contemporary American sport: A case study. *The Journal of the Christian Society for Kinesiology, Leisure, and Sport Studies* 2: 15–32.

Milbank, John, Catherine Pickstock, and Graham Ward, eds. 1999. *Radical Orthodoxy: A New Theology*. London: Routledge.

Millar, Robert. 2013. The Long Road to Redemption: Lance Armstrong Realizes It's OK to be Human. *CNN*. January 18. Available online: https://www.cnn.com/2013/01/18/sport/robert-millar-armstrong-oprah-cycling/index.html (accessed on 15 June 2019).

N.A. 2014. Armstrong Says Cycling is Still in a Mess after His Doping Confession. *Cycling News*. November 5. Available online: http://www.cyclingnews.com/news/armstrong-says-cycling-is-still-in-a-mess-after-his-doping-confession/ (accessed on 20 December 2014).

N.A. 2016. I Was a Complete Dickhead. *Newstalk*. October 7. Available online: https://www.newstalk.com/sport/i-was-a-complete-dickhead-lance-armstrong-off-the-ball-71405 (accessed on 20 October 2016).

Nelson, Dean. 2009. 'Holy' Moments Surround Us: You Don't Have to be Religious to Know That There's Something Bigger Out There, Often in Plain Sight. *USA Today*. October 26. Available online: http://deannelson.net/docs/usa_today.pdf (accessed on 30 June 2019).

Novak, Michael. 1992. The Natural Religion. In *Sport and Religion*. Edited by Shirl Hoffman. Champaign: Human Kinetics.

Oriard, Michael. 1982. *Dreaming of Heroes*. Chicago: Nelson-Hall.

Pelletier, Justin. 2019. Masters Champion Tiger Woods, Love Him or Hate Him, Good for the Game. *The Boston Globe*. April 14. Available online: https://www.bostonherald.com/2019/04/14/masters-champion-tiger-woods-love-him-or-hate-him-good-for-the-game/ (accessed on 20 April 2019).

Pucin, Diane. 2009. The Chase is on Again. *Los Angeles Times*. January 19. Available online: https://www.latimes.com/archives/la-xpm-2009-jan-19-sp-tour-lance-armstrong19-story.html (accessed on 10 February 2009).

Rapoport, Abby. 2013. Austin Loses its Hometown Hero. *The American Prospect*. January 18. Available online: https://prospect.org/article/austin-loses-its-hometown-hero (accessed on 16 August 2019).

Rishe, Patrick. 2012. Armstrong Will Lose $150 Million in Future Earnings after Nike and Other Sponsors Dump Him. *Forbes*. October 18. Available online: https://www.forbes.com/sites/prishe/2012/10/18/nike-proves-deadlier-than-cancer-as-armstrong-will-lose-150-million-in-future-earnings/#5f22cf766298 (accessed on 20 October 2012).

Roberts, Jimmy Jack McBee. 1982. Isaiah in Old Testament Theology. *Interpretation: A Journal of Bible and Theology* 36: 130–43. [CrossRef]

Rotunno, Tom. 2012. Armstrong Loses Eight Sponsors in a Day. *CNBC*. October 18. Available online: https://www.cnbc.com/id/49462583 (accessed on 20 October 2012).

Sanderson, Jimmy, and Marion E. Hambrick. 2016. Riding along with Lance Armstrong: Exploring Antapologia in Response to Athlete Adversity. *Journal of Sports Media* 11: 1–24. [CrossRef]

Sansone, David. 1988. *Greek Athletics and the Genesis of Sport*. Berkeley: University of California Press.

Scholes, Jeffrey, and Raphael Sassower. 2014. *Religion and Sport in American Culture*. London: Routledge.

Schrotenboer, Brent. 2016. Lance Armstrong's ban is partially lifted. *USA TODAY Sports*. September 7. Available online: https://www.usatoday.com/story/sports/cycling/2016/09/07/lance-armstrong-cycling-ban-partially-lifted/89981404/ (accessed on 17 September 2016).

Schrotenboer, Brent. 2018. Lance Armstrong Agrees to $5 million Settlement of Government Lawsuit. *CNBC*. April 20. Available online: https://www.cnbc.com/2018/04/20/lance-armstrong-agrees-to-5-million-settlement-of-government-lawsuit.html (accessed on 20 April 2018).

Sydnor, Synthia. 2003. The Radical Orthodoxy Project and Sport History. In *Transitions in Sport History: Continuity and Change in Sport History*. Edited by Wolfgang Buss and Arnd Kruger. Hanover: R. Kunz; Schriftenreihe des Niedersächsischen Instituts für Sportgeschichte, pp. 24–39.

USADA. 2012. Lance Armstrong Receives Lifetime Ban and Disqualification of Competitive Results for Doping Violations Stemming from His Involvement in the United States Postal Service Pro-Cycling Team Doping Conspiracy. *U.S. Anti-Doping Agency*. August 24. Available online: https://www.usada.org/lance-armstrong-receives-lifetime-ban-and-disqualification-of-competitive-results-for-doping-violations-stemming-from-his-involvement-in-the-united-states-postal-service-pro-cycling-team-doping-conspi/ (accessed on 24 August 2012).

Williams, Armstrong. 2009. Absurdity of Athlete Worship. *The Washington Times*. December 4. Available online: https://www.washingtontimes.com/news/2009/dec/14/the-absurdity-of-athlete-worship/ (accessed on 20 December 2009).

Wilson, Dave. 2019. How a small Texas town has reacted to the hiring of Art Briles. *ESPN.com*. July 11. Available online: https://www.espn.com/college-football/story/_/id/26943300/how-small-texas-town-reacted-hiring-art-briles (accessed on 14 July 2019).

Windsor, Richard. 2014. 'There is potential for redemption for Lance Armstrong' says Brian Cookson. *Cycling Weekly*. October 28. Available online: https://www.cyclingweekly.com/news/latest-news/potential-redemption-lance-armstrong-says-brian-cookson-141604 (accessed on 1 November 2014).

religions

MDPI

Article

Sporting Space, Sacred Space: A Theology of Sporting Place

Robert Ellis

Regent's Park College and the Faculty of Theology and Religion, University of Oxford, Regent's Park College, Pusey Street, Oxford OX1 2LB, UK; robert.ellis@regents.ox.ac.uk

Received: 27 June 2019; Accepted: 7 August 2019; Published: 10 August 2019

check for
updates

Abstract: Religion often designates locations that are considered sacred, marked off from ordinary space. Sporting venues also take on a significance for players and supporters that is seldom adequately explained in solely sporting terms. Can theological understandings of place illuminate the way in which players and spectators relate to the 'sacred space' of their sporting endeavors? In this paper, I explore and assess the theological and religious significance of sporting space by reflecting upon descriptions of both religious and sporting special places. I use a range of types of descriptions of experiences of such spaces together with theological ideas and concepts, including Christian notions of incarnation, sacrament, and Trinity, which are found to be useful resources, undermining a strict binary of 'sacred' and 'profane' space. I then build upon previous theological and empirical work with sports participants to explore a theological understanding of special sporting places and the experiences of those who play and support sporting endeavors in them.

Keywords: place; sacred space; religion and sport; theology and sport; sacrament

1. Introduction

Religion often designates 'special' places—locations that are considered sacred and are marked off from ordinary space. Ninian Smart's seminal work includes such places among the 'material' dimension of religion (Smart 1998, pp. 22–24). For Smart, the material dimension of religion encompasses triptychs, crucifixes, and fonts, as well as the elements of oil, bread, wine, and water deployed in sacramental observance. It also includes complex sites, such as the Ganges or Jerusalem, as well as locations such as white-washed dissenting chapels and baroque Cathedrals. What each of these material artifacts has in common is not mere religious utility but a capacity to carry and convey meaning and to mediate the experiences to which they witness. Furthermore, whether formally or informally, these artifacts have been made, or are regarded as, 'special'—separated or set apart, 'holy', or 'sacred'.

Sporting venues also take on a complex significance for groups of players and supporters that is seldom adequately explained in solely sporting terms and that—*prima facie*—seems similar in certain respects to that of special religious places. Some sporting locations are invested with iconic significance and have national and international resonance: one thinks, for instance, of Augusta's golf course or Lord's cricket ground. However, even local venues may take on the nature of 'hallowed turf'. Some sports clubs present their venues as sites for quasi-religious practices, with ashes of former players or spectators scattered or memorial events being staged there. Others offer less formal but rhetorically developed understandings (see Liverpool FC 2018).

Can the notion of 'sacred space' help us understand special sporting places and can it further help to explain the apparently (quasi-)religious experiences that sports participants often have in these locations? In this paper, I explore and assess the theological and religious significance of sporting space and raise the question for future work of whether such an understanding of sporting places sheds light back upon religious locations.

I begin by offering descriptions of experiences of two Christian and two sporting special places—a process that will allow some initial similarities to become obvious before I critically consider theological notions of 'place'. The Christian doctrines of incarnation, sacrament, and Trinity are deployed to this end. In this paper, a particular focus will be placed upon the notion of sacrament, but this, in turn, draws upon Christian ideas of incarnation. Additionally, the doctrine of the Trinity is useful, in particular, because it underlines the way in which God may be affirmed as present and active in the world while still allowing for clear differentiation between God and the world. These three doctrines undermine a strict binary between the 'sacred' and the 'profane', but the designation of especially holy places ('sacred space') is still useful. I then build upon previous work with players and supporters to explore the way in which sporting places may be experienced by analogy with religious 'places' and how both serve to mediate the experiences of the communities that gather within them. Particularly important for the theological concepts discussed here is the idea of self-transcendence and its grounding in the human person, who is understood to be made in the image of God. Insofar as sporting places may be described as sacred and sporting experiences may be described as religious experiences, I suggest that human self-transcendence is close to the heart of the matter. However, given the scope of this essay, it will not be possible to explore these aspects in detail (though I have addressed them elsewhere): the focus here is on 'place'.

A word must be said about the vocabulary of 'space' and 'place'. In popular conversation, the expression 'sacred space' is used frequently. However, academic discourse increasingly draws a distinction between 'space' and 'place'. The difference is nicely expressed by theologian Oliver O'Donovan:

> "Place" differs from "space," in that space is prior to culture and inhabitation, whereas place is the way we come to experience space when we have made our home in it. A "place," therefore, is the fruit of civilization, an area of space that has been distinguished from other areas by the inhabitation of a community. (O'Donovan 1989, p. 46)

Some writers see notions of 'space' as ideologically fraught, attempting to suggest a neutrality—a void in which things may or may not happen; and thus, the physicality of our actual space is diminished (e.g., Bartholomew 2011; Inge 2003). By contrast, 'place' is storied and the locus of embodied existence. As O'Donovan goes on to stress, a 'place' need not have a current population but is nevertheless construed by the human relationship to it. We understand our natural world, including those parts of it where the human imprint is faintest, according to a complex set of cultural factors (Sheldrake 2001, p. 16). Nevertheless, in general:

> [T]o think of a place is at once to think of a natural space on the one hand and of the community that is defined in relation to it on the other. It is to grasp the reciprocal relationship between nature and culture; geographical space mediating a possibility for human life in community; human inhabitation elevating dead space into the character and distinctness of place. (O'Donovan 1989, p. 47)

In this essay, I tend to speak about 'place' rather than 'space' but also use the term 'sacred space', recognizing that this expression is part of our vernacular.

2. Special Places

What can the notion of 'sacred space' mean to us today? It will perhaps direct us to religious buildings and sites where nature and religion seem symbiotically connected or to some other location where the descriptor 'sacred' might jar or surprise. My first two examples of special Christian sites are not comprehensive, but they are representative and suggestive. They show the interplay of environment and history with experience as well as the ambiguities of such places. From them, we will turn to two examples of special places in sport.

2.1. Sites of Celtic Pilgrimage

Twenty years ago I took a Celtic tour, visiting the Holy Island of Lindisfarne off the northeast coast of England and Iona off the west coast of Scotland. I went curious but skeptical. Brought up in the radical Protestant Baptist tradition, I was not naturally inclined toward the belief that places in and of themselves hold some kind of special aura.

Both destinations were bleakly beautiful islands with an untamed liminal feel, and the geography of both helps to create their sense of apartness. To both, a committed and careful journey is necessary. Perhaps it was the lingering morning mist; the solitariness of the location; the oddness of the tidal island's geography; the connection with the Celtic spirituality resources originating there (e.g., Adam 2010); or the long history of devotion, mission, and prayer; however, in ways I could not quite articulate, the Holy Island felt different. Somehow, 'spiritual experience' felt more possible.

On Iona, I stayed with a group of strangers in a community house. George McLeod, the Iona community's founder, was reported to have said that Iona is a 'thin place, where the veil between things spiritual and things material is as thin as gossamer' (Sheldrake 1995, p. 7; Power 2006, p. 45). McLeod was creative in his development of what one might call 'Iona mythology' (Power 2006, pp. 41–45), but his language of the 'thin place' has proved resonant and durable. While Iona bustled with visitors, it was still possible to feel something of what McLeod may have meant. The strong sense of community that built through the week was nourished by the rhythm of worship and a sense of common values and purpose. The narrative stretches back to Columba in 563 CE and was appropriated by the Iona community, the iconic abbey with visitors washing in and out like the tide, the island's remarkable light and ever-changing weather, and its physical isolation and beauty, all gave, similar to Holy Island, a strange sense of nature and history interweaving. One morning, we sang 'Morning has broken' in the abbey while birds sang in the building's rafters, which seemed to aptly sum up the experience.

Laura Béres' accounts of three visits to Iona are found in two overlapping articles (Béres 2012a, 2012b). She reflects upon her own experience and conversations with fellow pilgrims and residents, as well as upon the island's history and geomorphology. Béres explores some of this complexity with the help of postmodern geographers, social work theorists, and writers on spirituality. Her definition of spirituality is 'the human quest for personal meaning and mutually fulfilling relationships among people, the non-human environment, and, for some, God' (Béres 2012b, p. 183. The definition is Canda's; Béres cites Zapf 2009). A key element in this definition is the inclusion of both relational and environmental aspects, and Béres draws a contrast between the very different approaches to land or place in, for example, autochthonous traditions and Eurocentric ones. The process in the West of what Weber called the 'disenchantment of the world' (Weber 1964) may be thought to be challenged by what is regarded as an ancient Celtic interest in the significance of places—an interest that appears to survive even a skeptical reading of the recent Celtic 'revival'. Béres finds that geography and spirituality are bound together in Iona, reporting that one visitor claims that 'Celtic spirituality also is in tune with the elements and the patterns and rhythm of nature and that Iona is a place in which you can be more aware of these rhythms' (Béres 2012a, p. 400).

Each of her visits produces more complex and richer reflection. Initially, it is mostly the pilgrims in view. She reports that their experience is quickened by the journey to Iona and that a sense of community nurtures spiritual awareness. One of her interviewees reported that 'each time she came to Iona she was able to listen to God better and learn through the constantly changing physical scene around her' (Béres 2012a, p. 399). The importance of both physical surroundings and community is significant in this person's sense of the specialness of place. In her diary, Béres herself describes a contrast between this 'thin place' and the 'thicker' place from which she had come:

> Iona felt like home. It felt as though the saints and angels were walking with me … I felt completely safe. And here, back in Glasgow with the Orange parade going on, it feels as though there is a thickness … a buffer of distraction and human failing, with layers and layers of stuff and junk; a feeling of thickness rather than thinness. (Béres 2012a, p. 400)

On subsequent visits, Béres notices other things—particularly Iona's permanent residents, of whom there are just over a hundred. She muses on the socio-political relationships and competing Christian traditions and begins to ask whether the island 'seen' by tourists and pilgrims might be an idealized version of reality: might the residents feel colonized, marginalized, and exploited by the island's religious economy? Geographers and anthropologists describe places as sites of power struggle, resistance, annexation, and absorption (Inge 2003, p. 27). What Béres finds among the residents is not only pragmatism and some bemusement about a 'sentimental' attachment to Celtic ways but also some appreciation of the benefits of visitors and the Community (Béres 2012a, p. 402). Alongside a view of Iona as a special place, a much more complex picture of the island emerges: layers of experience and perspective, as well as history, nature, economics, and social relationships. The island does not 'mean' the same thing for everyone.

2.2. Cathedrals

Durham Cathedral is built on a site chosen by monks from Lindisfarne in the late tenth century. The current building is an impressive Norman edifice begun around a hundred years later. It has a fine collection of stained glass, and the remarkable rose window dapples the stone floor with color, while the vaulted ceiling strains toward the heavens. When I visited, I paused at the chapel dedicated to a local infantry regiment, musing on its appropriateness, and then at the shrine to Cuthbert of Lindisfarne and the tomb of the chronicler Bede. As I sat quietly in the nave, the silver cross on the altar caught the light. Amid the gentle echoes, there was a sense of peace and wonder.

Many readers will have had similar experiences in such buildings, with their blend of the universal and local. Perhaps we recall the remarkable light flooding through the lantern above the central crossing at Ely, the stunning modern stained glass and Sutherland's Christ in Majesty in Coventry, the numinous intimacy of St. David's, the contemporary space window and Darth Vader gargoyle in Washington, D.C., or the awesome, organic wonder of the still-growing La Sagrada Familia in Barcelona. Such places commonly evoke silence, reverence, and awe, connecting, in a mysterious way, those present with something other worldly.

Systematic theologians begin their *magna opera* in various ways. Graham Ward begins his by describing his entry into Christ Church Cathedral in Oxford. 'The air is cool, the shadows deep and in the silence distant sounds reverberate' (Ward 2016, p. 3). By referencing 'deep shadows' and 'unheard sounds', Ward guides his readers to this sense of awe and mystery commonly reported in such magnificent buildings and prepares his readers for an apophatic strain in what will follow. Ward's cathedral is a different sort of space from the city's ordinary streets or its distinctive covered market:

> I tread more softly; my pace is slower; and my posture bends or cranes according to what I am seeing and trying to understand. Although I have been here a thousand times I can't domesticate the place ... it is a holy space; a space within which I am taught that name, that God, that who or what or operation circulating among so many of my other more mundane beliefs. (Ward 2016, p. 3)

He goes on as follows:

> It's not just the religious symbolism—crosses, an octagonal font, and the vast eagle across whose back is stretched a heavy Bible. It's not just that prayer has been said here for over a millennium. There is some inherent quality to this place, difficult to define, that makes it holy, that generates and requires holy practices that make us holy ... This is a place invested with care and value and attentiveness where people have wept and petitioned and fumbled for words and delighted. (Ward 2016, pp. 3–4)

In contrast to Iona and Lindisfarne, the physical environment that interacts with history here is primarily a human construction, and Ward observes that the holy practices of the cathedral connect worshippers with their Anglo-Saxon, medieval, and reformation forbearers, as each generation has shaped the space. It is a reminder that, as some of Béres' geographers put it, places always have a temporal as well as a spatial dimension (Béres 2012b, p. 178; referring to Massey 2009, p. 59); indeed, places without a temporal narrative are often characterized as 'wilderness' (Inge 2003, p. 38; Brueggemann 1978, p. 40). Ward notes some details of this space: the vaulted ceiling; the quality of the light falling on the sandstone pillars; the rich coloration of stained glass and clergy vestment; the aroma of incense and candle; the surfaces of brass, silver, and wood; the textures of the fabric; the impressive surge of the organ and the delicacy of choir song; the cruciform shape of the whole; and the altar and its cross. The cathedral has its own pedagogy, he observes: architecture and practices alike teach us what it is to believe in the Father, Son, and Holy Spirit.

However, the 'seams' of temporality in the space are not its only complexities. Ward draws our attention to dissonance as well as harmony and to the multiple and multifarious motivations that have produced the artifact in which Christians now worship (in this regard, see Brown 2004, p. 23). The temptation is to treat it as one morally unambiguous and coherent space—timeless and without history; however, it is a much more layered and complex reality. In this, the cathedral not only serves to open up the ground of a lived theology but also resembles that theology, which he says is always earthed, embodied, allusive, and oriented to worship but also flawed, self-interested, and ideological (Ward 2016, pp. 34, 144).

Space, even here, in the most obviously 'holy' of places, is neither devoid of human input nor lacking in ambiguity. It may be God's space, but it is also human space. Despite its ambiguities, the place evokes and nourishes an experience of the reality to which it testifies.

2.3. The Millennium Stadium

I meandered through the streets in the heart of the city with one of my sons and an old friend with whom I shared this walk decades ago as schoolboys. This time my son was sharing the 'pilgrimage' with us. We were walking into a shared past, and we recalled episodes from it as we headed toward the Millennium Stadium. This 'cathedral' is built on the site of Cardiff Arms Park, where Wales has played international rugby games for generations. Bleddyn, Barry, JPR, and Gareth—the names of the 'saints' resonate in Wales and in rugby, though they may be meaningless elsewhere. The stadium itself is similar to many other modern sports arenas. The turf is carefully laid and manicured, its emerald surface adds a touch of 'nature' to the concrete stands, and the grass alone sometimes produces gasps of wonder as fans emerge from the tunneled staircases. Nearby runs the River Taff, adding elements to the distinctive location of the stadium and its mythology. The sense of continuity with the past is strong.

This stadium was barely twenty years old, but in it, we were connected to past wonders ('on the site of'). It felt special. Inside, the crowd sang fragments of hymns and poured out devotions to their team; to today's chosen ones (chosen similar to priests representing communicants); to some idea of Wales—of 'us'; and to some connection with and valorization of a certain past—fact and fiction. That day, Wales beat Italy handsomely, and the singing continued. 'Bread of heaven, bread of heaven, feed me 'til I want no more . . . ' The experience in this place creates an identity that lives outside it, but what happens here nourishes and maintains it.

Yet, not everyone in Cardiff that day tingled with the occasion. For some, this was an opportunity to make some money as seventy-five thousand people flooded into the city; for others, the crowds were a nuisance and the game an irrelevance. Even the faithful were divided about who should be playing and especially about how their game—the nation's game—was organized away from this match: the 'cathedral' flourished, but the 'parishes' were in a mess.

Poet and novelist Owen Sheers became the first 'writer in residence' of any national rugby team when he took up that position with Wales in January 2012. His poem *Now and Then* (Sheers 2012) was published in the matchday program for Wales' game against France—the game Wales had to win in order to secure a rare grand slam by beating all the other teams in the tournament.

Sheers captures something of the 'meaning' of the Millennium Stadium for Welsh rugby when he speaks of the way in which 'a nation becomes a stadium ... moments when the many, through the few, become one'. The experience of 'Welshness' and its connection to rugby and to this place goes some way toward explaining why the stadium is experienced as special, separated. Space and time flow together, community is invoked and imagined, heroic acts and selfless sacrifices are recalled, something similar to worship and awe is felt, and the whole seems apart from ordinary reality. There is delight in beauty and grace as well as in power and determination. There are moments of surprise and suspense and of quiet and expectancy. Symbols of Wales are scattered everywhere. Sheers suggests that the whole 'land' is made somehow sacred by its association with this playing field in the center of Cardiff, deploying a cross-linguistic pun—*pridd* is Welsh for 'soil' and just a letter away from 'pride'. Those who watch are constituted by their shared action: 'A nation watching, sharing a pulse/as the clock counts up to the final whistle'. Wales' relationship to this place may be distinctive, but to fans of other sports, much will sound familiar.

2.4. West Ham United: Upton Park and the London Stadium

A further sporting example makes the point through negative contrast—the case of West Ham United FC and its stadium move in 2016. Once again, issues of history and physical space are interwoven.

West Ham had played at Upton Park since 1904, but the stadium had become increasingly unsuitable. After a controversial process, the team won the right to become tenants at the London Stadium, which is a few miles from Upton Park and was built for track and field events for the 2012 Olympics. However, being tenants with various restrictions has limited the ways in which they can make it 'feel like home'. An arena built for track and field is very different from Upton Park, where stands closed intimately and intimidatingly around the pitch. At the London Stadium, supporters behind the goals (where many diehard fans prefer to be) feel detached from the action, and sideline distances are still big. The stadium was 'converted' for football use through a system whereby the front tier seats were moved forward over the running track, creating a gap between the front and rear seats, increasing spectators' sense of isolation from the action and one another. Instead of feeling like a familiar (if slightly dilapidated) family home, the London Stadium is said to be a soulless corporate venue.

I interviewed lifelong West Ham fan ('Hammer') about the move. Hammer's parents had grown up in the West Ham area, and Hammer had attended games regularly since the age of 8. The Upton Park he remembered changed a lot over the years. Views were obscured by pillars, but crowd acoustics were good. He recalled the stadium as 'a bit shabby', even tacky at times. Notwithstanding, he said, 'it was home', and, significantly, 'they are mystical places, right?'

He recalled great wins, dismal failures, and shared match days with school friends, college friends, work colleagues, and family. He built new friendships with those who sat around him regularly. Hammer cherished what he called the 'match day ritual': parking in a similar place, going to the same café, and seeing other regulars do the same. This sense of the occasion 'beyond the stadium' is noted by others: 'I used to go to Upton Park, grab a program, nip in the pie and mash, have a bet, into the boozer, meet my pals, all good, have a laugh, then out afterwards [to the match]' (Anthony 2018). This sense of the significance of the journey to the stadium echoes dimly the pilgrims' experience and is in marked contrast with the experience of fans at the London Stadium. According to Hammer, the journey to London Stadium is neither as easy, as was promised, nor as convivial. He complains of feeling 'managed' through traffic systems in a characterless location. Moreover, there are problems with seating and remoteness, poor crowd acoustics, and the very visible presence of tourists in other clubs' shirts, which all add to problems of poor team performance, making the match day experience border on toxic—setting fans not only against the club's board but also sometimes against one another. The lack of any visual continuity with Upton Park serves to underline further the break: neither the statues of famous players nor the quiet grove in which fans' ashes were scattered—nor anything else, for that matter—was brought to, or replicated in, this new place.

Ward spoke of the pedagogy of the cathedral, and stadia too have their own pedagogies through which fans are educated in the tradition of the community associated with that special place. A fan arriving at Manchester United's Old Trafford is taught about the greatness of their club in various ways. A Spurs fan at their newly rebuilt stadium has visual cues as well. At the London Stadium, there is no pedagogy.

Hammer believes that a poor start in results has delayed the opportunity to build up new memories of memorable performances or to develop the balanced way in which fans mull over defeat and disappointment. Over time, these things will come, he thinks, though it is difficult to ignore an undercurrent of regret in his words, suggesting that things will never be the same as they were.

Anthony's article tells a similar story (see also Russell 2018; Stone 2017), reporting that the new stadium is 'soulless', and referring to Upton Park as the club's 'spiritual home'. Anthony observes that 'what has upset the fans this season is not the loss of a football match so much as the loss of an identity' (Anthony 2018). Another fan complains, 'we have given up our history, our heritage, our legacy, all of which was focused on Upton Park'. Such claims may seem overblown to those who are not invested in sports; however, places that bear a history create a people. In the interweaving of physicality and narrative and symbols and memories, places become special, 'sacred'. They evoke and nourish certain kinds of experience. This is not to deny an element of nostalgia. Upton Park moments are recalled with some rose-tinting. As it was becoming unsuitable for modern football, the area around it was changing rapidly as well, such that ' . . . Upton Park represented a link to a disappeared past . . . ' (Anthony 2018).

Anthony demonstrates, similar to Béres and Ward in different ways, the layered nature of special places, both sporting and ecclesial, and the ambiguities that are connected with such places when we tease away at the layers. What we find, again, is that 'real' places present many facets for examination. Christian places of worship and theologies are earthed, embodied, allusive, and oriented to worship, as well as flawed, self-interested, and ideological. They are sometimes constructions of our imaginations, which 'edit out' aspects of their reality. In similar ways, our other special places also present their ambiguities, if we choose to look and see; yet, these ambiguities do not rob them of their 'specialness'.

3. Theological Resources for Place and 'Sacred Space'

3.1. Initial Considerations

Protestant theologians have tended to be wary of ascribing any particular significance to special places, and if they do, they tend to insist that such specialness is simply a function of the people who have become associated with these places (see White 1995; Walker 1990, 1992). It was often argued that the New Testament (NT) spiritualizes the Old Testament (OT) approach to the land, and thereby to place generally, and that Christianity replaces the OT concern with place with a universally available risen Jesus (Davies 1974). No*where* is special anymore.

The work of Walter Brueggemann has prompted a re-evaluation of this thesis. Brueggemann argues that a special place, specifically the land, is the Bible's central theme. He speaks of 'place' as a location

> [W]hich has historical meanings, where some things have happened which are now remembered and which provide continuity and identity across generations [where] important words have been spoken which have established identity, defined vocation and envisioned destiny [and] in which vows have been exchanged, promises have been made, and demands have been issued . . . (Brueggemann 1978, p. 5)

In recent years, some Protestant traditions have begun to take the concept of special places more seriously (the Iona and Northumbrian communities are examples), and some Evangelical writers have similarly shown themselves to be more open to the subject. Tom Wright, for instance, speaks of his 'slow turning away from various forms of dualism, to which Evangelicalism is particularly prone, and towards the recognition of the sacramental quality of God's whole created world' (Wright

1999, p. 4). He recalls attending a concert in a school building that had previously been a church, in which he had a strong sense of 'the presence of God, gentle but very strong', noting that he 'had no theology by which to explain why a redundant United church should feel that way' (Wright 1999, p. 4). Sociologist E.V. Walker's remark that 'human experience makes a place, but a place lives in its own way' (Walker 1988, p. 2) perhaps hints at something similar. But how can we account for this theologically? As Wright acknowledges, until recently, it was usually Catholic writers who offered the theology he lacked to account for the specialness of places, which was usually based on the doctrines of incarnation or sacrament; however, this has begun to change.

3.2. Incarnation

The incarnation, it is typically argued, suggests that God's concern with reaching out to humankind is not merely universal, for every time and place, but engages particularity. The incarnate Word lives a human, emplaced life—in a particular culture and period and in a sequence of particular places. Sheehy insists that 'this appears to be a principle of God's working with us' (Sheehy 2007, p. 16). He quotes Thomas Torrance: 'While the incarnation does not mean that God is limited by space and time, it asserts the reality of space and time for God in the actuality of His relations with us, and at the same time binds us to space and time in all our relations with Him' (Torrance 1969, p. 67, cited Sheehy 2007, p. 19). William Temple's claim that Christianity may be said to be the most materialistic of the major world faiths (Temple 1934) is often cited. Matter exists in space, and if God expresses Godself through material, then God may also express Godself in place.

Human emplacedness is key here:

> To be human is to be placed: to be born in this house, hospital, stable (according to Luke), or even, as in the floods in Mozambique in 2000, in a tree. It is to live in this council house, semi-detached, tower block, farmhouse, mansion. It is to go to school through these streets or lanes, to play in this alley, park, garden; to shop in this market, that mall; to work in this factory, mine, office, farm. These facts are banal, but they form the fabric of our everyday lives, structuring our memories, determining our attitudes. (Gorringe 2002, p. 1)

3.3. Sacrament

The materiality of a sacrament is clearly an associated notion. As with incarnation, it suggests a divine choice and promise to communicate and mediate grace through material means. The historical development of ideas of sacrament has been controversial, but in the twentieth century, a creative renewal of theological thinking has led to the notion of Christ as the sacrament of God (Schillebeeckx 1963) and, following Vatican II, the idea of the Church as sacrament and even 'the brother as sacrament' (Balthasar 1958, pp. 142–55). A number of thinkers now speak of a 'sacramental universe' (Temple 1934; Macquarrie 1997; Sheldrake 2001; Bartholomew 2011). This may sound similar to the Quaker view that all of life is sacramental, but we should add an important qualification. To affirm the possibility that God may mediate his gracious love through a material thing is not to suggest that there is anything automatic about this—a view that might be thought to lead towards pantheism (Inge 2003, pp. 66–68). Rather, to speak of the sacramental is to speak of an event and, in particular, of a divine action. These are 'the rents in the opacity of history where God's concrete engagement to change the world becomes visible' (Gorringe 2002, p. 165)—language that recalls McLeod's talk of 'thin places'. A place, viewed sacramentally, is a material location through which God mediates saving presence and grace. To claim this in a meaningful way is to claim that the place is experienced in this way repeatedly—or 'reliably', as one might say—while recognizing the divine initiative and eschewing any suggestion of something quasi-magical.

As Brown suggests, Christianity's wariness about giving spiritual worth to 'ordinary places' stems from a tendency to downplay the value of the world and a concern that 'human beings might themselves set the conditions under which God could be experienced' (Brown 2004, p. 21). He observes that this objection might be made about the Church's sacraments as well, and this, in turn, leads us to suggest that the hesitation about 'ordinary places' might be occasioned less by the possibility that we could control the conditions under which we experience God than a concern that we cannot control them—or, more to the point, that the Church cannot control them. A sacramental understanding of place comes up against our ideologies and self-interest.

3.4. Trinity

More recently, Gorringe and Bartholomew have been among those suggesting that a Trinitarian approach can yield a positive theology of place. Gorringe, proposing a theology of the built environment, argues that all space, and thus every place, is *'potentially* sacred, waiting for the moment of encounter in which it mediates God' (Gorringe 2002, p. 40). In every place, as at Bethesda, an angel can stir the waters (John 5:4). A theological justification for this can be found in Barth's emphasis on the freedom of God: while God is present in every place, this is 'not uniform but distinct and differentiated', such that God is 'free to be present in some places in a way which God is not in others' (Barth 1957, pp. 473–74, cited by Gorringe 2002, pp. 42–43).

While a fallen creation means that God's gifts are distorted and his presence is never known without ambiguity, fallenness does not mean that there is no goodness in the world or that God is not active in it or cannot be mediated through it (Gorringe 2002, p. 15). The doctrine of the Trinity is considered key to navigating the ambiguity of our experience of the world and our hope to discern God's presence within it, offering, as current theological language puts it, the grammar for speaking about God in relationship with the world and suggesting that God is both present to and different from the world. The doctrine of the Holy Spirit helps believers to find God in life, freedom, creativity, and vitality and in the forging of relationships and community. The incarnation provides a hermeneutical key with which to approach scripture and our present experience. However, the doctrine of creation *ex nihilo* insists on God's absolute difference from the world.

Thus, the doctrine of the Trinity works dialectically, in Lash's words, 'to indicate where God is to be found—and by denying, at each point, that what we find there is to be simply identified with God—to prevent us from getting stuck in one sidedness … The doctrine thus leads at every turn, to both affirmation and denial' (Lash 1988, p. 267).

3.5. Place as Sacrament

Bringing these ideas together, we can frame a theological question regarding place. Given that God chooses particularity in the incarnation, promises to give Godself sacramentally through material, and is free to choose this place rather than that place, could it be possible that God would choose some particular places, such that they appear to evoke a sense of God's presence in and of themselves? A positive answer may appear to marginalize or discount any human element and underestimate the importance of what happens in a given place as opposed to the 'place itself'. However, the 'place itself' is an unhelpfully abstract idea. We have already noted that any place is understood through the lenses of human culture and experience and the history of human interaction (or the lack thereof) with it. One cannot understand a place without also considering its historied aspects. As indicated, definitions of 'place' combine nature and culture. Even when place is not cultivated, the way we experience 'wild' places will be shaped by cultural factors. Theological understandings of place recognize this, with evangelicals insisting on the specialness of a place residing in the people who have been associated with it and sacramentalists arguing that the specialness of place lies in the history of divine disclosure there (Inge 2003, p. 79)—complementary positions. Thus, Inge argues for a relational view of sacred places that triangulates God, people, and locations (Inge 2003, pp. 46–47, 78–89), giving greater prominence to the material location but linking it indissolubly with its community as well as divine action. Such a view needs refining and qualifying to explain both the 'new place' (such as a new church building—a place with no history of divine encounter) and the 'old place' (such as the former church building, where the people who had shared the history of divine encounter are no longer present to keep the memory alive; it is as if the very stones somehow held the memories), but it represents a strong building block in the argument for special places.

3.6. Sacred and Profane?

Vernacular talk of a 'sacred space' appears to imply a binary between that which is and is not sacred. There is a widely accepted distinction between the shrine and the shop or changing room, which is often underscored by Durkheim's seminal treatment of the sacred (Durkheim 1915, pp. 38–39) and reinforced by Eliade's insistence on the 'non-homogeneity of space' (Eliade 1959). While this *prima facie* appears compelling, the universalizing tendency that W.D. Davies (Davies 1974) and others propose that Christianity introduces may not so much do away with all separated or sacred space as relativize the distinction between it and other places. Nevertheless, every religious tradition maintains some form of special place—some form of the sacred; and unless certain locations are designated 'holy' in some distinct way, we may not know how 'holiness' appears in other locations (see Hegdes 2018, pp. 12–18). 'God is omnipresent but that does not mean that his presence can therefore be felt everywhere equally . . . [and] foci help . . . ' (Brown 2004, p. 386).

The hard distinction maintained by both religious and sociological/anthropological discourse is not one that can be accepted straightforwardly by Christian theologians, as our consideration of the above doctrines makes clear. We must insist that while some places are separated off in their special sacredness, we should not miss the potential of every place to be holy ground—as Moses found in the desert, confronted by the burning bush (Exodus 3). We should resist any tendency to separate 'sacred space' too rigorously and find a balance between the rejection of the notion of sacred space and the banal affirmation that every space is always sacred. There are special places, but every place has the potential to become the vehicle of divine encounter. Thus, it may be possible to affirm the following:

> Earth's crammed with heaven,
>
> And every common bush afire with God . . . (Browning 1857)

Every place is God's place (Psalm 24:1)—every bush can flame with God's presence. What the above-considered doctrines suggest is that this is neither automatic nor controllable by human agency. God determines, in divine freedom and grace, where and when God acts and is known. Nevertheless, the divine promise, also given in freedom and grace, appears to commit to (we might say 'covenant' with) humankind to encounter us at particular locations and in particular ways. These 'places', we have seen, are complex, ambiguous, and historied sites, always defined in some respect by human interaction with them. Some are 'obviously' special, established for the purpose of communing with God, whereas others appear more accidentally so. While we may feel close to God in a cathedral, we might have similar experiences in other locations as well (Gorringe 2002, p. 37). Sporting locations might be counted among the other places that might be or become special.

Tillich's theology of culture is useful in conceptualizing this relativization. For Tillich, religion and culture are not different spheres occupying, as it were, separate spaces within the cultural landscape—sacred and profane. Rather, the religious is present, potentially, in every aspect of life as the 'dimension of depth', which directs our attention to 'that which is ultimate, infinite, unconditional in man's [sic] spiritual life' (Tillich 1975, p. 7). In every 'preliminary concern, ultimate concern is present, consecrating them' (Tillich 1975, p. 41). Thus, every place, even those not designed or designated as 'sacred', can open up to us this dimension of depth. Sport is a cultural activity in which the dimension of depth may open up, and sporting locations may mediate this limpidity.

4. Sporting Places as Sacramental Locations

Bain-Selbo asserts 'that human beings act religiously in ways other than those restricted to institutionalized religion, . . . that the sacred is not restricted to one's church, and that spiritual communities can be formed outside of stereotypically religious congregations' (Bain-Selbo 2009, p. 232).

We now consider sporting spaces as material locations in which people 'act religiously' and form some kind of 'spiritual community' by reflecting upon their sacramental character.

The sacramental character of special places should not be restricted to 'natural' places, nor specifically to 'religious buildings' (in affirming this, Brown also considers homes, gardens, and whole cities: Brown 2004, p. 245). A brief consideration of the ways in which Brown argues that church architecture conveys and elicits meaning (their 'pedagogy') will be helpful. His study concentrates on the ways in which buildings focus a sense of transcendence or immanence or both. Key to the former is often the way in which light is used by architects to suggest an 'other worldly' light and also to draw congregants towards it worshipfully. Similarly, and most obvious in Gothic styles, the use of vertical lines in columns, towers, and spires draws the eye upwards and 'Godwards'. Both techniques rely on deeply primitive symbols for transcendence—height and light. Immanence may seem less apparent, but Brown believes that the tendency in modern church design to gather the community around the altar (or the music group, perhaps) in ways that emphasize community is one way of doing this. This over-simplified sketch offers just enough to continue our analysis. In what follows, I will be thinking mainly, though not exclusively, of major venues and stadia as shared 'special' places of sport.

Approaching a modern stadium, one cannot but be struck by its considerable bulk. Concrete and sometimes glass rise from the ground impressively. We are used to large buildings, and from the outside, modern stadia are often not conventionally beautiful—though seeing the whole, or more of the whole, can add to the aesthetic pleasure (as in the aerial views so popular now on television, showing stadia in their aesthetically satisfying circular shapes). However, even by the standards of large buildings, the Millennium Stadium is big. Its massive bulk stretches up before us. Whether it is appropriate to speak of awe in such an approach remains to be seen, but some of the raw materials are there, including some of those verticals—though the more rounded, Colosseum-like shape of these structures softens them. The architects of the new Wembley Stadium incorporated allusions to the old Wembley through the new arch, expressing continuity with the past through an immediately recognizable motif. Reaching out beyond and above, straining at or toward something, the arch has practical as well as symbolic use (Wembley Stadium 2019; Tekla 2019), and similar to many a flying buttress, it looks very impressive. The pilgrims travelling up Wembley Way take all this in—the primitive symbols impacting them unconsciously—and some kind of impression is made, stirring a response.

There comes a point in the approach to a modern stadium when suddenly a view opens up of the field. Through the concrete, the light suddenly changes, and we glimpse the rich emerald of the playing surface. Many fans will feel a frisson of excitement at this view—a combination of the change in light (accentuated if there are floodlights) and the sudden burst of color amid the drab greys and beiges. This 'wow' moment can be followed by others when emerging into the stands or terrace, including the play of light on sporting uniforms. There is continuity between Brown's account of transcendence-evoking architecture in churches and the design and experience of sport's special places (if we were to consider, in contrast, golf's special places, we might need to consider Brown's analysis of cultivated gardens or depictions of nature in art to seek other clues to transcendence and immanence). Players and supporters alike report aesthetic satisfaction and enjoyment in sport. Often, these reports relate to the play rather than the arena, but not always. With the color, the light, and the sights and smells, all the fans' senses are engaged and stimulated in the stadium. This involved and involving experience demands and gives energy.

There is also immanence. Old stadia were often constructed and developed piecemeal: a terrace here; a stand there. Aligned around each side separately, they were often an exercise in getting as many people into small spaces as easily (though not as safely) as possible. Some supporters mourn the loss of the old standing terraces. The physicality of supporting a team often feels cramped by a requirement to stay seated: it is easier, perhaps, to kick and tackle with one's team's players when standing, as well as to gesticulate, sway, and generally engage in 'bodily worship'. Standing in older stadia, we literally 'felt' part of a community. The atmosphere in such spaces was a key part of their popularity, and this atmospheric aspect of the stadium has not disappeared in modern grounds but has, rather, been transmuted (though we recall the problematic London Stadium above). Typically, in modern stadia, the seating wraps around the pitch in a bowl-like shape, giving a feeling not unlike the congregation gathered around the altar. In Britain, a roof may keep the sound in and amplify the noise, enhancing the aural dimensions of the immersive experience. The uniformity of the seating, the noise of support, the orchestration of crowd response, and often the proximity to the playing surface all reinforce the community feel of the experience. There is a dimension of religious experience that may be categorized as 'communal' (Ellis 2014, pp. 179, 185–86), and it is close to aspects of what Brown describes in speaking of an experience of the immanence of God.

In the special sporting place marked by history, memory, words and deeds, and promises and demands, a 'nation'—or a town, community, or group of disparate individuals—'becomes a stadium', and a common sense of identity is created. Players and spectators speak of the important social bonds of special sporting places. Often these are trans-generational, connecting the living to the dead: the frequent requests for the scattering of ashes in stadia or memorial gardens adjacent to them testify to this. However, in addition to this temporal dimension that draws upon and adds to the storied layers of the experience, there is the spatial dimension that connects individuals together in the present: supporters report that a 'feeling of belonging' is generated and speak of the 'supportive family' of their clubs (Ellis 2014, p. 170). For supporters, presence at the game, at a place, is about identity and belonging (see Gorringe 2002, pp. 73–76) but also about seeking certain kinds of experience in the reciprocal relationship between players, supporters, and location. The 'sung liturgies' of sport relate to identity and identification: they used to be affirmed with team scarves and rattles but are now done so with 'replica shirts'. The impact of consumerism on these phenomena does not eclipse the reality that they project. We recall that a Trinitarian approach anticipates the ambiguities of the world. God may be mediated through such experiences without being fully identified within them. In a similar way, while the purity of the sporting experience may be debased by various factors, the reality of the experience itself remains.

The immersiveness of the experience propels some supporters and players towards those experiences of 'flow' that have so often been compared to mystical experiences (Ellis 2014, pp. 155–456, 179–80), and it appears that the material location of the stadium plays a part in this as in the generally 'religious' character of the in-stadium experience.

Modern stadia have a tendency to resemble one another in ways not always true of the oddities of old, and it may be thought that they lack charm or individuality. Some supporters think of them as 'soulless', and this sense of alienation from place is most pronounced when combined with an unwanted move away from a more familiar 'home'. The experience of Cardiff Blues' rugby supporters after the team's short-lived move to Cardiff City's soccer stadium adds testimony to our London Stadium stories. A more extreme example of exile is that of the move and renaming of Wimbledon FC to Milton Keynes in 2003. There is no sense of continuity with the old, only a feeling of betrayal and loss, deeply felt, similar to something from the Psalms.

The special places of sport are not only its great 'cathedrals' or pilgrimage destinations but may also be particular and local places associated with remembered events and relationships, longing and promise, and community and identity: large stadia and local playing fields. These places may be understood to function sacramentally in the ways in which they mediate the experience of sporting participants.

5. Assessing the Experience of Special Sporting Places

We now approach, at first obliquely, a question posed at the outset. Can the notion of 'sacred space' not only help us understand special sporting places but also further help to explain the apparently (quasi-)religious experiences that sports participants often report in these locations?

I first came across Dorothy Gurney's poem 'God's Garden' when I read its closing couplet inscribed around the edge of a garden pond in the home of a devout Christian.

One is nearer God's heart in the garden

Than anywhere else on earth. (Gurney 1934, p. 9)

The gardener was a woman with a sense of place. She had an attachment to her own local church's building that was more than mere sentiment, and she took enormous pleasure and real spiritual nourishment from the natural world. She did not feel that the 'religious experience' that she had in her garden was one that belonged to a parallel realm to the experience that she had in church; it was not, as it were, a similar affective response to different 'things'. She affirmed that it was the God and Father of her Lord Jesus Christ in both cases (or perhaps the Risen Christ who met Mary in the guise of a gardener). In this, she echoed the experience of those who have often been called 'nature mystics'. Rather than worshipping divine nature, as is sometimes alleged, these nature mystics claim to have an experience of God through the materiality of nature.

We considered earlier Brueggemann's account of what 'makes' place: it is marked by historic meanings or memories of the past that provide continuity and identity through generations, and it is a location where important words and deeds have been spoken and done and where vows are made, promises given, and demands issued—all of which establishes identity, vocation, and a sense of destiny. Does this sound like a way of describing sporting sacred places? I suggest that it does rather a good job of doing so for the Millennium Stadium, for Lord's cricket ground, and for many others, including local venues. The complex interplay of factors proposed by Brueggemann illuminates some of the multifaceted sinews that bind supporters to players and clubs or teams. Here, we see how 'when the many, through the few, become one', a sense of identity is created and reinforced. Pilgrimage, itself having many layers, is driven, among other factors, by the creation and nurturing of identity, and the concept of pilgrimage proves useful for explicating significant sporting journeys. A lifelong cricket fan, when I attended my first Lord's test match a few years ago, the experience had much in common, albeit in a modest way, with a pilgrimage. Brown is correct to resist restricting 'pilgrimage' to 'religious' destinations (Brown 2004, pp. 153–54).

In our special places, that which is 'bigger' than the individual—identity, community, memory, and meaning—coalesces, construed as expressed in and carried by a location and its various elements. This is as true in a cathedral as in a sporting venue. There it is God, who Christians believe is known and experienced. What of the stadium? At the very least, it seems that the special places of sport function in a way that is analogous to the ways in which special places in religion function. A similar dynamic of community, memory, and location is at work, and the places generate their own ways of mediating and nurturing the experience. But what is this experience? Is what players and supporters sometimes report in the special places of sport an experience of some parallel but phony realm, or is it an experience of the same God known by Christians in Jesus and (sometimes) experienced in worship, prayer, and discipleship; in the sacramental materiality of bread and wine; or in the sacramental love of Christian fellowship? Is the experience known in playing or supporting a genuine experience of the transcendent, however diffracted or distorted? For while it may be appreciated that God may be experienced through the world (indeed, how else could God be experienced?), Christian theologians have also regarded it as a fallen world, a place of ambiguity, and have considered that God, who is experienced through the material is not identified wholly with it.

The experience of sports, both as a spectator (vicarious sport) and player, is an experience that is congruent with the richness of those experiences that we label 'religious'. In particular, sport offers its participants experiences that might be described as mystical, charismatic, regenerative, communal, and nurturing (Ellis 2014, pp. 179–89). The special places in which sport is played and celebrated contribute in various ways to these experiences, evoking and mediating them in a sacramental way. The desire to win, to improve one's performance, and to transcend the present self is at the heart of sport for all its participants (Ellis 2014, pp. 128, 233–48). This, too, is mediated sacramentally by place, which teaches, shapes, and draws out this desire, and it is the centrality of this self-transcendence that suggests that sporting experience may not be merely a substitute for religious experience but, rather, the thing itself. The restless desire at the heart of sport may be understood by reference to Rahner's description of human openness toward God and marked by a similar sense of restless desire. Rooted in human consciousness, human persons are always questioning and probing at their limits, always reaching beyond themselves, an indication that we are 'always still on the way. Every goal that [we] can point to in knowledge and in action is always relativized, is always a provisional step. Every answer is always just the beginning of a new question' (Rahner 2010, p. 34). For Rahner, every restless reaching out is ultimately a quest for God, the ground of all our probing. Seen in this light, sporting experience may have more in common with religious experience than being a mere simulacrum.

It would seem reasonable to suggest that a 'religious experience' that is self-conscious is thus in some way fuller or more complete; however, in reflecting upon the similarities and/or differences between experiences in the stadium and (for instance) those in church, it may be helpful to consider the way in which C.S. Lewis compares the values encountered in great literature with the values of the Christian faith. He regards these values as 'sub-Christian', though this expression should not be understood in the derogatory way in which one might use the term now. Instead, he suggests that, while they are something less than what Lewis regards as fully Christian values, they represent the highest level that can be attained without attaining the values of the Christian faith. He speaks of them as an anticipation (a kind of starter or *hors d'oeuvre*) or reflection (as the moon to the sun) of Christianity. He says, 'though "like is not the same," it is better than unlike. Imitation may pass into initiation. For some it is a good beginning. For others it is not; culture is not everyone's road into Jerusalem, and for some it is a road out' (Lewis 1981, p. 40).

Seeing a stadium experience as an anticipation of a fuller and more fully aware kind of religious experience may be regarded as a useful way of comparing the two phenomena. However, Lewis' observation that not everyone will 'travel to Jerusalem' through an experience of culture—or of sport—is also helpful. As Lewis suggests, it may lead in another direction altogether, because the striving for self-transcendence that lies at the heart of sport may also become disordered (Ellis 2014, pp. 262–74). Some of the more problematic aspects of sport may form its participants in undesirable ways, and a distorted competitiveness is as likely to lead away from as toward God. However, the difficulties of sport do not eclipse its positive possibility. As Lewis also says, 'imitation may pass into initiation'.

When Brown finally comes to consider sport, it is disappointing that he only does so as a metaphor for religious experience in modern cinema. However, he indicates elsewhere that we might properly encompass sport and sporting places in a view of a world that is alive with the potential presence and action of God:

> The natural world, the layout of a town or garden, the structure of a specific building, a basketball shot can all induce religious feelings that ought not to be summarily dismissed as though necessarily inferior to a Christian's experience of response to prayer or of worship in a church. God can be encountered in both types of experience alike as a given, the former exhibiting the same non-instrumental character that worship ought rightfully to have. (Brown 2004, pp. 407–8)

I submit that the foregoing suggests that, insofar as sport can be a vehicle for an encounter with the divine, it is in no small part because of the role played by its special places in mediating such experiences.

Funding: This research received no external funding.

Conflicts of Interest: The author declares no conflict of interest.

References

Adam, David. 2010. *Tides and Seasons: Modern Prayers in the Celtic Tradition*. London: SPCK.

Anthony, Andrew. 2018. 'It's Soulless Here.' Why West Ham Fans are in Revolt. *The Guardian*. April 29. Available online: https://www.theguardian.com/football/2018/apr/29/why-west-ham-fans-are-in-revolt (accessed on 7 August 2019).

Bain-Selbo, Eric. 2009. *Game Day and God: Football, Faith, and Politics in the American South*. Macon: Mercer University Press.

Balthasar, Hans Urs Von. 1958. *Science, Religion, and Christianity*. Translated by Hilda Graef. London: Burns and Oates.

Barth, Karl. 1957. *Church Dogmatics. Volume II, Part 1*. Translated by Geoffrey Bromiley. Edinbugh: T & T Clark.

Bartholomew, Craig. 2011. *Where Mortals Dwell: A Christian View of Place Today*. Grand Rapids: Baker.

Béres, Laura. 2012a. A Thin Place: Narratives of Space and Place, Celtic Spirituality and Meaning. *Journal of Religion & Spirituality in Social Work: Social Thought* 31: 394–413. [CrossRef]

Béres, Laura. 2012b. Celtic Spirituality and Postmodern Geography. *Journal for the Study of Spirituality* 2: 170–85. [CrossRef]

Brown, David. 2004. *God and the Enchantment of Place*. Oxford: OUP.

Browning, Elizabeth Barrett. 1857. *Aurora Leigh*. London: Chapman & Hall.

Brueggemann, Walter. 1978. *The Land: Place as Gift, Promise and Challenge in Biblical Faith*. London: SPCK.

Davies, William David. 1974. *The Gospel and the Land: Early Christianity and Jewish Territorial Doctrine*. Berkeley: University of California Press.

Durkheim, Emil. 1915. *The Elementary Forms of the Religious Life*. Translated by Joseph Ward Swain. London: Allen & Unwin.

Eliade, Mircea. 1959. *The Sacred and the Profane*. New York: Harcourt & Brace.

Ellis, Robert. 2014. *The Games People Play: Theology, Religion, and Sport*. Eugene: Wipf & Stock.

Gorringe, Timothy. 2002. *A Theology of the Built Environment: Justice, Empowerment, Redemption*. Cambridge: CUP.

Gurney, Dorothy Frances. 1934. God's Garden. In *God's Garden and Other Verses*. London: Burns, Oates & Washbourne.

Hegdes, Paul. 2018. Deconstructing Religion: Some Thoughts on Where We Go from Here—A Hermeneutical Proposal. *Exchange* 47: 5–24.

Inge, John. 2003. *A Christian Theology of Place*. Farnham: Ashgate.

Lash, Nicholas. 1988. *Easter in Ordinary: Reflections on Human Experience and the Knowledge of God*. London: SCM.

Lewis, C. S. 1981. Christianity and Culture. In *C. S. Lewis: Christian Reflections*. Edited by Walter Hooper. Glasgow: Fount.

Liverpool FC. 2018. We are Liverpool: This Means More. Available online: https://www.youtube.com/watch?v=YhqiSO_UFxg (accessed on 7 August 2019).

Macquarrie, John. 1997. *A Guide to the Sacraments*. London: SCM.

Massey, Doreen. 2009. *For Space*. London: Sage Publications.

O'Donovan, Oliver. 1989. The loss of a sense of place. *Irish Theological Quarterly* 55: 39–58. [CrossRef]

Power, Rosemary. 2006. A Place of Community: "Celtic" Iona and Institutional Religion. *Folklore* 117: 33–53. [CrossRef]

Rahner, Karl. 2010. *Foundations of Christian Faith*. New York: Crossroads.

Russell, Michael. 2018. What's wrong with West Ham? A lifelong fan explains . . . . *GQ*. March 12. Available online: https://www.gq-magazine.co.uk/article/whats-wrong-with-west-ham-a-lifelong-fan-explains (accessed on 7 August 2019).

Schillebeeckx, Edward. 1963. *Christ, the Sacrament of*. Translated by Paul Barrett. London: Sheed and Ward.

Sheehy, Jeremy. 2007. Sacred Space and the Incarnation. In *Sacred Space: House of God, Gate of Heaven*. Edited by Philip North and John North. London: Continuum.

Sheers, Owen. 2012. Now and Then. Available online: https://www.facebook.com/owensheersauthor/posts/owens-poem-for-the-france-programmenow-and-thenwhat-might-have-been-and-what-has/10150599249280614/ (accessed on 7 August 2019).

Sheldrake, Philip. 1995. *Living between Worlds: Place and Journey in Celtic Spirituality*. London: Darton, Longman and Todd.

Sheldrake, Philip. 2001. *Spaces for the Sacred: Place, Memory, and Identity*. Baltimore: John Hopkins.

Smart, Ninian. 1998. *The World's Religions*, 2nd ed. Cambridge: CUP.

Stone, Simon. 2017. West Ham: Why has London Stadium move been so problematic. *BBC Sport*. September 12. Available online: https://www.bbc.co.uk/sport/football/42329101 (accessed on 7 August 2019).

Tekla, Trimble. 2019. Available online: https://www.tekla.com/uk/references/wembley-stadium-arching-ambition (accessed on 7 August 2019).

Temple, William. 1934. *Nature, Man and God*. London: MacMillan.

Tillich, Paul. 1975. *Theology of Culture*. London: OUP.

Torrance, Thomas. 1969. *Space, Time and Incarnation*. New York: OUP.

Walker, Eugene Victor. 1988. *Placeways: A Theory of the Human Environment*. Chapel Hill: University of North Carolina Press.

Walker, Peter W. L. 1990. *Holy City, Holy Places? Attitudes to Jerusalem and the Holy Land in the Fourth Century*. Oxford: Clarendon.

Walker, Peter W. L., ed. 1992. *Jerusalem Past and Present in the Purposes of God*. Carlisle: Paternoster.

Ward, Graham. 2016. *How the Light Gets In: Ethical Life I*. Oxford: OUP.

Weber, Max. 1964. *The Sociology of Religion*. Boston: Beacon Press.

Wembley Stadium. 2019. Available online: http://www.wembleystadium.com/about/press/stadium-facts-and-features (accessed on 7 August 2019).

White, Susan. 1995. The Theology of Sacred Space. In *The Sense of the Sacramental*. Edited by Ann Loades and David Brown. London: SPCK.

Wright, Tom. 1999. *The Way of the Lord*. London: SPCK Triangle.

Zapf, Michael Kim. 2009. *Social Work and the Environment: Understanding People and Place*. Toronto: Canadian Scholars' Press.

religions

MDPI

Article

Pray the White Way: Religious Expression in the NFL in Black and White

Jeffrey Scholes

Department of Philosophy, University of Colorado Colorado Springs, Colorado Springs, CO 80918, USA; jscholes@uccs.edu

Received: 6 July 2019; Accepted: 31 July 2019; Published: 6 August 2019

check for
updates

Abstract: Athletes, particularly players in the National Football League, have repeatedly invoked God in order to glorify, praise, or even credit the divine with success on the field. This essay examines the ways in which different types of religious language used to bring God onto the gridiron are received and evaluated along racial lines. I seek to show that speech by athletes, in particular black football players, that communicates a God who is partisan and intervenes in action on the field is routinely dismissed by fellow players, the media, and religious authorities who favor a God who either intervenes softly and generally or is above the game altogether. I contend that a double standard is applied to this theological debate due to a disregard of historical African American theology and to hegemonic white evangelical norms that police such discourse.

Keywords: race; national football league; religious expression; prayer; black church; providentialism; evangelicalism

1. Introduction

David J. Leonard, in his book *Playing While White*, compares and contrasts the ways that trash-talking in sports is digested and interpreted along racial lines. He finds that white athletes who trash-talk (belittling the opposition to gain an advantage), such as Tom Brady, Peyton Manning, Larry Bird, and Michael Phelps, are praised by the media and fans for this behavior. The talk demonstrates passion and fierce competitiveness rather than a petty or even mean comportment. Alternatively, the trash-talk dished out by black athletes such as Muhammad Ali, Allen Iverson, Serena Williams, and Shannon Sharpe is often deemed disrespectful, uncivil, and a sign of societal decline.

Leonard calls out this glaring double standard for what it is—the amount of melanin in one's skin determines how your trash-talk is received. It is within a protected space that white athletes move and speak, and therefore their trash-talk is at worst "childish" but at best "illegible" for its inability to be translated as truly disrespectful despite the disrespectful language used (Leonard 2017, p. 73). Within the relatively unprotected space that black athletes occupy, where blackness is equated with "criminality", "rage", and harmful hip-hop culture, trash-talking somehow furnishes us with a window into the character of the athletes (Leonard 2017, pp. 66, 69, 84). And that character is putatively damaging the sport and setting a poor example for young fans.

As Leonard chronicles, white quarterback Johnny Manziel during his sophomore year at Texas A&M pretended to give an autograph to an opposing player after converting a first down and then flashed a "money-counting gesture" throughout the game and in subsequent games. This behavior was taken by some media members as a sign of his "gladiator mentality" and only an indication that Manziel was just "having fun" (Leonard 2017, pp. 76–77). Alternatively, for black cornerback Richard Sherman, the tongue-lashing that he administered on opponent Michael Crabtree after preventing a touchdown received the complete opposite reaction. By some in the media, Sherman was called a thug; was immature, classless and dangerous. His speech-act was characterized as an unprofessional tirade and garbage (Leonard 2017, pp. 82–83). Leonard interprets this and other similar situations that he cites as such,

> White America is deeply uncomfortable with African American trash-talkers; the sight and sound of black emotion, whether anger or joy, whether frustration or passion, elicits ample reaction, dissection, and commentary. Bravado and confidence, like rage, is unacceptable in association with blackness. Whiteness plus brashness is not only acceptable but also desired and celebrated. (Leonard 2017, p. 84)

Because there is "no mention of the cultural practices of signifying . . . no discussion of oral traditions or cultures of resistance" (Leonard 2017, p. 74) that infuse the African American experience and its expression, trash-talk uttered by black athletes is left exposed to appropriation that is largely guided by stereotypes that are applied to blacks and whites with equal measure. Manziel's "whiteness contributed to an alternative reading of his words and his trash-talking swagger", whereas Sherman's language, "reflects the danger seen in black bodies that lack 'discipline, that are not under control'" (Leonard 2017, pp. 75, 86).

I start this essay with the double standard applied to trash-talking to set up a question: is a similar double standard applied to the religious expressions of athletes? Religious expressions such as prayers, statements of thanks, religious gestures, etc., are also considered to be extraneous to sport. These expressions have likewise raised the ire of the media and fans throughout the years. And these kinds of statements or gestures can be bold and brash. I contend that the racialized dual-interpretation of trash-talking is analogous to that of the treatment of religious expressions in the NFL. This double standard is predicated on the assumption that most religious expressions, no matter the race of the player, rely upon the premise that God is active in the world and hence involved in the game of football in some way. The difference between God giving strength to players and giving victories to players is minimal theologically, but apparently vast culturally.

Like trash-talk (or any talk), the religious expressions of professional football players and coaches emerge from cultural contexts that help clarify and even justify the content and style of the expression, whether spoken by white or black players. Yet, the voicing of the conviction that God does intervene in the sport itself, whether held by black or white athletes, is too often abstracted from the historical and political contexts from which they arise respectively often rewarded or conveniently overlooked when uttered by white athletes and dismissed as nonsense or labeled as heterodox or worse when spoken by African Americans with no consideration of the sources

After offering a typology of religious language in the sporting world, I examine the historical and cultural sources that black and white religious expressions rely on. I then turn to several examples of African American football players who use what I call "hard providential" language to articulate their understanding of their play on the field and the critical responses to such language made by white players, media members, and religious authorities. Finally, I look at the way that white evangelical quarterback Tim Tebow relates God to football through a "soft providentialism" that calls on God to give him strength and keep the game injury-free—a theology that enjoys scant notice. Tebow's religious stance is, in part, protected by a double standard that is encouraged by long-standing ideas about race and sustained by white evangelical norms. It can be upheld only when historical and political circumstances that have molded religious expression are disregarded.

2. Types of Religious Expression in the NFL

It is estimated that around 40% of the members of teams in professional baseball, basketball, and football in the United State are evangelicals (Krattenmaker 2010, p. 13). And in a 2016 survey of the players on the NFL's Denver Broncos, 86% identified as Christian and 90.7% of those surveyed said that religion "was very important" to them (Klemko 2017). In addition, one-fourth of Americans believe that God directly affects football games, and over half of Americans say that God "rewards athletes who have faith with good health and success" (Cox and Jones 2017). Religious adherence may be waning amongst millennials overall in the United States (The Age Gap in Religion around the World 2018), but not so in the sports ranks, particularly with football players and its fans.

There exists a wide range of the types of religious expression in the NFL. Each expression, whether vocal or not, is performative—these athletes and coaches are on a big stage and each utterance is meant to have impact. In addition these expressions have a theology behind them as well, albeit usually unacknowledged nor well developed. A laying out of a brief typology will not only help with a kind of theological categorization but also with the later building of the hierarchical structure that arranges and slots each type into its place along racial lines.

Of course religious expression, like any expression, is the product of complex, often murky sources. Add in the performative component and motive, style, and setting further muddy the water. Moreover, these categories are far from rigid. Many players draw on language found in more than one category—sometimes within the same sentence. Yet as we will see, some types of religious expressions overlap considerably because the theologies and cultural mores undergirding them overlap considerably as well. Where differences are patently evident, it behooves us to ask about the theology within and the authority it is given by a certain culture that makes the expression resonant and/or legitimate with some and not with others.

The sequence of the following typology moves from expressions that convey a God that very lightly intervenes (if at all) in sport to a God who directly interferes with action on the field. The first type, the "it's just a game" type of religious expression, dictates that some things matter much more than sport. When religion is invoked to drive this home, the theology behind it is that God is bigger than football and therefore cares not about play on the field or the outcome. This type of expression does not have to be lodged in a religious context, however. Tony Romo used a version of the phrase after a particularly devastating loss, perhaps as a defensive measure that would minimize the psychological toll. "If this is the worst thing that will ever happen to me, then I've lived a pretty good life" (Curtis 2017).

But for other football players more inclined to locate their game and life in a heavenly context, this kind of statement not only lifts God above the fray, thus safeguarding God's impassibility but also acts as a means of overcoming tough losses, career-ending injuries, and being traded or cut. And if refuge is taken in an impassible divine figure, the troubling vagaries of the game can be relativized in the face of a God who is distanced from sport altogether. This type of religious expression is put nicely by New England Patriots' special teamer, Matthew Slater:

> This is not life and death. This is not the biggest thing that we'll do. The biggest and most important decision we'll ever make in our lives is what we do with the knowledge of Jesus Christ. And if you don't know the answer to that question it can be challenging at times. But I think for people that look at this game as the end all, be all, it's not! It's a game. We're thankful to be here, we're thankful for the opportunity. We love what we do but we have to keep perspective on the greater good! (Buehring 2019)

The "billboard" type of religious expression involves God in the game more than the previous type but still relieves the athlete from having to defend the role that God may play in sport. It entails the mere posting of a bible verse or a pithy religious message on a uniform or on social media without commentary. Gestures such as pointing to the sky, making the sign of the cross, kneeling or genuflecting, as quarterback Tim Tebow popularized with his "Tebowing", or "Praise performances", as Shirl Hoffman dubs all of them (Hoffman 1992, p. 157), are functional equivalents to the presentations of a Bible verse. The intent behind them leaves the audience to decipher their meaning. Some of these billboards are hidden but accessible to a few such as the beginning words of Philippians 4:13, "I can do all things ... ", stamped on the inside of the tongue of basketball star Stephen Curry's name-brand shoe.[1] The NFL restricts this kind of messaging far more stringently than does the NBA,[2] though in 2017, it began to allow approved messages to be shown on players' shoes for a designated week during the season. For example, Philadelphia Eagles' quarterback, Carson Wentz chose to emblazon "AO1" (Audience of One) and "Romans 5:8" in gold caps across the bottom of his cleats.

These standalone displays are largely provocative. After Tebow wore "John 3:16" on his eye black in the 2009 college football championship game, ninety-three million people were provoked to Google the verse. But because the Bible says little about athletic activity and nothing about football, drawing conclusions about the theology of Tebow and why he advertised the verse in this way, aside from the content of the verse, is difficult. One can hide behind the billboard. It will do all of the talking.

The "glorification" type of religious expression centers on individual or team action on and off the field that glorifies God. Almost exclusively inspired by 1 Corinthians 10:31, "So whether you eat or drink or whatever you do, do it all for the glory of God", athletic activity is included in the category of "whatever you do." The act of glorification is performed by an athlete or a coach, but by reflecting God's greatness in the glorifying process, the expression is meant to "give God the glory", not the athlete. These actions are often tangential to the outcome of a game and more tied to how the game is played. Did I play the game ethically? Was sportsmanship demonstrated? Was maximal effort expended? Am I leading by example rather than berating or belittling the opposition? Were emotions channeled in the right direction and therefore not made manifest in foul language or disrespecting the opponent or one's own teammates? If so, then God's qualities were mirrored, and God is thereby glorified in the process.

Sometimes the amount of glory given to God is thought to be amplified on a big stage after a win. Or God is given glory by the performance itself but making sure a large audience knows "the real reason" why one plays the game can bring God even more glory, as Dallas Cowboys' quarterback Roger Staubach iterated after winning the 1972 Super Bowl. "I had promised that it would be for God's honor and Glory, whether we won or lost. Of course the glory was better for God and me since we won, because the victory gave me a greater platform from which to speak" (Blazer 2015, p. 59).

Because it is the actions of the athletes that drive the glorification rhetoric, this type of expression can come off as a kind of humblebrag, as the Seattle Seahawks' quarterback Russell Wilson tweeted after signing a thirty-five million dollar-a-year contract, "My Hallelujah belongs to YOU" followed by the hashtag, #All For Your Glory. This kind of statement at once is intended to deflect attention away from his own role in securing the contract while the dollar amount shines the light back on Wilson, who is in the process of glorifying God through his success (Parke 2019). One way around this accusation is to emphasize that glorification occurs not through the behavior that reflects God's will, but rather that God's glory is made manifest through a receptive, though imperfect container. For athletes that may possess less raw talent than others, this "vessel" metaphor is more efficacious theologically and culturally than the "mirror."

The last two types of religious expressions involve a God who intervenes the game more directly. The first "soft providential" type and the second "hard providential" type loosely follow the Reformed distinction between general and special providence. General providence entails the overall upholding of creation by God for the good of its inhabitants. God watches over creation and governs it invisibly. Therefore, the effects of general providence are not usually felt as long as a sense of order and overall good is registered. In sports language, this kind of soft providence is relied upon with petitions for safety on the field, strength to persevere, and a sense of peace to counter the violent nature of the game. Denver Broncos defensive lineman Domata Piko states it thusly, "I always ask God before the game and I ask him to give me the strength that he gave to Samson to defeat the Philistines, and I say give me the courage that you gave Daniel when he faced the Lions. I want God's strength out

---

[1] Curry has been writing this verse on the outside of his shoe since college and has commented on the practice.
[2] Tebow used to write chapter and verse on his eye black under his eyes during his football-playing college career, but that practice is disallowed at the professional level.

there." For Piko, God is intervening softly by providing strength and courage that presumably would be lessened in the absence of this gift.

By definition, providence, whether hard or soft, involves an intervening God. According to the dictates of soft providence, an injury-free game may serve as evidence of the kind of divine intervention that is at work. Though if injuries occur, rarely is God's supposed non-intervention or failure to intervene called out. We can suppose that it is far more onerous to exonerate or credit God within the hard providential model, though. Athletes may publically praise a God who keeps them safe, strong, and faithful with impunity. If a freak injury happens, it is not typically theologized. If strength, resolve, and faith were not present on the field, it is rarely acknowledged as God's failure—it is on the athlete or team.

God is prevented from immersing Godself in the messiness of a football game within the soft providential category but not so in the hard providential mode. This second type of providence involves a God who cares who wins, alters play on the field, and affects outcomes through miraculous intervention. This is a partisan God, not unlike the God of the ancient Israelites, who works through athletes and coaches to achieve the desired end. There is much less distance between human and divine will here as athletes who apply the hard providential type to their expressions are claiming to divine God's will as evinced through plays on the field. Hall of Fame quarterback Kurt Warner communicates the difference between hard and soft providentialism nicely by rejecting the former in favor of the latter:

> When you stand up and say, "Thank You, Jesus", they think you are saying, "Thank You for being here. Thank You for moving my arm forward and making the ball go into that guy's hands so that we could score a touchdown and win the game." But, in essence, it is a matter of thanking Him for the opportunity, thanking Him for being there in my life, for being the stronghold, for being the focus and the strength to accomplish all things, to accomplish anything, and to be where I am at, to have gone through everything I have gone through.
>
> (Schuster 2014)

3. Religious Expression in Black and White

As noted earlier, Leonard problematizes the absence of the role that history, sociology, economics, and politics play in differential treatment of trash-talk delivered by black and white athletes. The same can be said of religious expressions. If the influence on the content and style of rhetoric are appreciated, then the single standard applied to blacks and whites but written and governed by white norms alone invites scrutiny.

The context for much religious expression emergent out of the African American community is slavery and the institutions established under of Jim Crow law. Slave religion, a syncretistic blend of indigenous African traditions and the slave owners' selective offerings of Christianity, needed to respond to the present oppressive conditions. A bound, obedient body and a muted, circumscribed spirit were required in public for survival under slavery. But a body that could be made free through the innervation of the Holy Spirit was a body that stood in opposition to its status on the plantation provided brief, though powerful, refuge.

The spirit's manifestation was corporeal and often wild. W.E.B. DuBois describes an early twentieth century Southern revival thusly:

> Finally the Frenzy of "Shouting", when the Spirit of the Lord passed by, and, seizing the devotee, made him mad with supernatural joy, was the last essential of Negro religion and the one more devoutly believed in all the rest ... And so firm a hold did it have on the Negro, that many generations firmly believed that without this visible manifestation of the God there could be no true communion with the Invisible. (Du Bois and Marable 1999, pp. 120–21)

While DuBois was somewhat skeptical of the metaphysical reality of the Holy Spirit and uncomfortable with the unrestrained gesticulations, Anthony Pinn argues that spirituality is crucial for his definition of the "black soul" as made visible through the body. "Possession by the holy gives the poverty stricken body new value, a new level of spiritual beauty that overrides—at least during the period of *communitas*—the physical realities of life under the *status quo*" (Pinn 2003, p. 4). Or according to Albert Raboteau, "Prayer, preaching, song, communal support, and especially 'feeling the spirit' refreshed the slaves and consoled them in their times of distress. By imagining their lives in the context of a different future they gained hope in the present" (Raboteau 1980, p. 218). On Pinn's reading of DuBois, "through the shout or frenzy black bodies were transported beyond the confinements of the Veil and placed in communion with that which affirmed their humanity" (Pinn 2003, p. 3). Pinn translates this historic black pneumatology for the contemporary black church.

> The contemporary Church seeks to establish blacks as agents of will–with all the accompanying benefits and responsibilities. Christian gatherings orchestrated by the church often serve as a ritual of exorcism so to speak in that they foster a break with status as will-less objects—impoverished forms—and encourage new forms of relationship and interaction premised upon black intentionality and worth. (Pinn 2003, p. 3)

Or the difference between being a slave, where the spirit was necessary to construct a new free body, and a legally free, but still oppressed African American citizen is that the spirit is galvanizing the will. The black body still endures the white gaze and bodily maltreatment, but the spirit, in conjunction with the will, enables the body's navigation and negotiation in a white world.

This, albeit brief and incomplete, description of the historical and social context of this strand of African American theology can assist us in making sense of the use of hard providential language by black athletes today. If God, whether understood as the Holy Spirit or otherwise, has been employed to reimagine the black body as free, willful, and responsible, it should come as no surprise that this understanding permeates the world of professional football, which is 70% black. With black bodies risking brain or bodily injury that could end a career (and a paycheck[3]) within a milieu of exclusive white ownership of teams and vast majority of white fans, the claim that the supernatural works its way through a natural body towards success exhibits an internal logic. When survival, both on the field and in economic terms, is determined by performance, it is no wonder that God is sometimes given credit for exemplary performance. Triumph is experienced as freedom not wholly unlike the kind experienced within the confines of the black church.

Whites have largely enjoyed the luxury of not having to call on God or the Holy Spirit to reconfigure their bodies in a way that enables the overcoming of societal imprisonment or deleterious stereotypes. God can be conveniently called on to maintain the status quo. Yes, whites, particularly in the South and Appalachian regions, have espoused a similar interpretation of the workings of the Holy Spirit and expressed it with bodily flailing and glossolalia (and have likewise been dismissed by the mainstream). Yet these groups possess the skin pigmentation that solicits judgement of their parochial, "backward" culture without the threat of political punishment for holding such beliefs. African Americans who hold to a relatively similar doctrine and exhibition are doubly punished.

The conventional means of maintaining the political status quo while remaining staunchly Christian for whites has meant trusting in God's soft providence. This way, the sustenance of the "worldly" state of affairs can be comfortably aligned with God's intervention. When that state of affairs is disrupted, it is usually the reflex of the people whose status is at risk to stifle the disruption by sidelining it rhetorically at best, removing it at worst. Granted, the means to preserve social order have changed over time as society has changed. Yet the habit of disciplining the speech, presentation, and actions of African Americans by the white majority dies hard. It is with this in mind that we turn

---

[3] Unlike the NBA, MLB, and NHL, the NFL does not guarantee the terms of any contract with a player. Injury or poor performance (or other considerations) is grounds for immediate severance of the contract with the player, no matter the length of time or amount of money "promised" under the original contract. See https://www.economist.com/the-economist-explains/2018/10/09/how-nfl-contracts-give-players-so-little-power.

to the several examples of such disciplinary tactics that are applied to language spoken by several African American football players that is of the hard providential type.

4. God Does Not Care about Football

On 18 January 2015, the Seattle Seahawks came from behind late in their game to beat the Green Bay Packers in the NFC Championship game. It was particularly stunning that the aforementioned Russell Wilson led the comeback to propel Seattle to its second straight Super Bowl after he had thrown four interceptions earlier in the game. Immediately after the game and on the field with a microphone in his face, Wilson relayed his version of the sequence of events. "That's God setting it up, to make it so dramatic, so rewarding, so special. I've been through a lot in life, and had some ups and downs. It's what's led me to this day." In the Green Bay locker room, the opposing quarterback, Aaron Rodgers, who is white and a devout Christian at the time[4], was told of Wilson's comments. Understandably upset over the outcome as well as the suggestion that God did this to his team, Rodgers replied, "I don't think God cares a whole lot about the outcome. He cares about the people involved, but I don't think He's a big football fan." When the two teams met up the following season and Green Bay exacted its revenge on the Seahawks, Rodgers snarkily remarked that "I think God was a Packer fan tonight, so he was taking care of us."

David French, in an article in the *National Review*, uses this back and forth along with other references to God made by athletes to challenge calls by some prominent journalists to get God out of sports. He states, "Rodgers's original distaste for Wilson's comments highlights an underacknowledged truth: the role of God in our personal and professional successes (and failures) is a matter of no small controversy ... " Rodgers' comments comminucate to French how seriously Christians, especially Christian football players, take their theological commitments. He asserts, "to get God out of football, the anti-religious crowd would need to get the football players out of football" (French 2015). French ends with a celebration of the spectrum of religiosity in football as he appreciates both, "Wilson's effusive declaration of his beliefs" and "Rodgers's more understated perspective" without comparing the theology of the two.

While French is likely correct in his assessment of the religious conviction held by a sizable majority of professional football players and hence the impossibility of its vanishing, he misses a crucial point. Rodgers won not only the second game but the rhetorical battle as well. Wilson's hard providential language, that God intervened in the championship game to effect the outcome, is met with a comeback that is more generally accepted in American culture. God has more important things to do than pull the strings on a sporting event, especially as it concerns strings that deliberately cause mistakes only to unveil God's ultimate handiwork. Rodgers then delivers the knockout blow with a sarcastic mimicry of Wilson's interventionism to point out its inconsistency and ludicrousness.

Journalist Dave Zirin is not so equanimous. "When it comes to football, there is no separation between God and the public square. Hearing Aaron Rodgers give a little nationally televised pushback against the idea of a higher power being deeply invested in a football game is as satisfying as it is overdue" (Zirin 2015). Significant for my purposes that I will address shortly, is that neither commentator mentions the race of either quarterback nor its importance in the rhetorical battle.

Another case that displays this kind of back and forth is that of Reggie White, Hall of Fame defensive lineman and ordained minister who passed away in 2004. White has voiced his share of proclamations about God having a direct hand in his career and his play on the field. The "Minister of Defense" certainly talked about glorifying God with his play and used soft providential theology at times to relate God to the game he played. Yet White also spoke as frequently of a partisan God

---

4   In an interview given in 2017, Rodgers, when he still identified himself as a Christian, replied that he does not affiliate with any religion. http://www.espn.com/espn/feature/story/_/page/enterpriseRodgers/green-bay-packers-qb-aaron-rodgers-unmasked-searching.

who picked his teams to win and intervened to secure the bet. "God intervened in David's fight over Goliath and in Jesus' victory over death", so, White reasoned, how can the football field be off-limits?

White has also credited God with stepping in and guiding his career decisions. He became a free agent in 1993 after playing with the Philadelphia Eagles and repeatedly told the media that God would tell him which team to play for next. He chose Green Bay and recounted the dialogue as such.

> I thought, I know God told me to go to San Francisco. What's the deal? And the Lord spoke to me. And when the Lord spoke to me, he said "Let me ask you a question: Where did the head coach, the defensive coordinator and the offensive coordinator all come from before they went to Green Bay?" I said, 'San Francisco?" And he said, "That's the San Francisco I'm talking about!" (Howard 1996)

The disappointed and skeptical Eagles' owner Norman Braman said after White signed with the Packers that Reggie's decision, "wasn't going to be made by a ghetto or by God. It was going to be made for the reason most human beings make decisions today: money." White clapped back, "I just thought, how dare Mr. Braman say that? Money was important, because I needed resources to continue the projects I wanted to do. But how dare he speak for what was in my heart? He doesn't know me. We had dinner. But he never walked down any streets with Reggie White" (Howard 1996). Braman, who is white, attempts to discredit White's story by accusing him of confusion at best (Reggie may think God is talking to him, but it is really Mammon) and lying at worst.

Adding insult to injury, the press had fun with the rumor that it was not God who told White to join the Packers but then coach Mike Holmgren impersonating God. According to the story that Holmgren swears is true, the coach, knowing who the real authority in Reggie's life was, left this message on his answering machine. "Reggie, this is God. Come to Green Bay." Reggie signed with the Packers a short while later. Holmgren's tactic, while clever and effective, betrays a condescension for White's personal belief that God intervenes in the world and speaks to him. That Holmgren could trick White into signing with the Packers by replacing himself for God reveals what he (and the media) thinks of White's theology and its susceptibility for exploitation.

Former Baltimore Ravens' linebacker and Hall of Famer Ray Lewis has similarly relied on hard providential logic. Lewis has unapologetically proclaimed that God does indeed take sides in a football game. When asked, "How does it feel to be a Super Bowl Champion?" Lewis responded, "When God is for you, who can be against you?" This lone statement prompted an Episcopal priest to write an essay disputing Lewis' logic and theology. According to the priest, Lewis is preaching the "Gospel According to Ray without consequence or accountability" that skirts a "fine line between God and Ray … that not only makes people uncomfortable, it can be dangerous" (Schneck 2013). The danger is cast in theological terms: if there is equation between any human (besides Jesus) and God, then God is subject to the desire of individual humans. God's agency cannot be dragged down and steered in this way.

Lewis' Hall of Fame speech in 2018 faced similar denunciation. Lewis proclaimed that he is God's messenger, that the number he wore (#52) was God's number, that God healed him of a nasty injury, and that God spoke to him and calmed him while he was dealing with an accusation that he was involved in a double murder.[5] One journalist called the speech "meandering and mostly meaningless" (Van Bibber 2018). Another called it "frothing madness" and a "desperate, sweaty, rapturous self-exaltation" (Thompson 2018). The thrust of these critiques is that Lewis preaches what God has done for him, but this only runs cover for Ray to talk about himself. Yes, hard providential language suffers little daylight between the divine and its instrument. This can open the speaker up to allegations of self-promotion and egocentrism from secular critics; idolatry and blasphemy from the

---

[5] Lewis was indicted on a murder charge in 2000 for the stabbing deaths of two people that he and two of his friends were seen fighting with. Lewis was given a misdemeanor plea in exchange for pleading guilty to obstructing justice with misleading statements the night of murder. The case remains unsolved.

religious sort. That Lewis believes that God intervenes in all aspects of his life, including his play on the field, is somehow proof that he is unhinged, narcissistic, and therefore not to be taken seriously.

In one final example, the headline on the cover of Sport Illustrated's 1998 pre-Super Bowl issue read, "Does God Care Who Wins the Super Bowl?" Provocative and outside of the magazine's usual questions posed to its readers, this seemed a good one to ask after hearing from some of the players about to take the field in the upcoming game. Several players from the Denver Broncos, all black with one exception, viewed their win over the Pittsburgh Steelers to earn a trip to the Super Bowl as an act of God. Fullback Howard Griffith elaborated on an important play that he made in the game: "I attribute everything to Him. The Lord allowed me to make that catch." Further, "God directs everything, He already knows the outcome. Kordell (the Steelers' quarterback) didn't go out there to throw those interceptions. The game was already decided before we walked out there." Tight end Dwayne Carswell reflected on two crucial interceptions made by the Broncos: "God could have caused that. He's in everybody's corner, but I guess He decided that we deserved [to win]." And tackle Tony Jones answered the magazine's opening question with a "yes", God cares specifically about the Broncos. "He's been looking out for us the whole season. We've been through some tough storms, but He brought us through. Now we are on our way to San Diego, and we know He is with us." To no surprise, Reggie White, who was playing for the upcoming opposing Packers, opined, "God had a lot to do with this, and I praise God that I had a chance to win one Super Bowl last year and now another."

Interestingly, the article leads with these comments by black athletes and is led the rest of the way by white players, theologians, and religious leaders who proceed to undercut what came before. This second part is set up with loaded questions that goad readers into formalizing expected answers.

> Does God take an active interest in the outcome of athletic matches? Did He favor Denver over Pittsburgh or Green Bay over San Francisco? Does a believer on one side of the ball have an advantage over a nonbeliever on the other side of it? Does God even know there is a Super Bowl?

Adam Timmerman, white guard for Green Bay, is given first crack at an answer. "I ask him to keep us from injuries. And I ask for victories: 'God, I want to win so I have an even bigger platform to use for you.' People listen to winners more than they do to losers."

With that from inside the locker room, the article moves outside the lines. Richard J. Wood, dean of the Yale Divinity School, approaches the issue by separating divine omniscience from divine interest. "It doesn't seem to me odd that God would know in detail what happens in football games. What seems to me odd is that God would care." Given the "more important" events going on in the world, "that God has a direct involvement in athletic contests trivializes the whole notion of God's involvement with the world. It is a heresy." Joseph C. Hough, dean of the Vanderbilt Divinity School, damns the thought of God sinking down to the level of football games. It "makes God look immoral and arbitrary. I find that religiously offensive." And former president of Fuller Theological Seminary, Richard Mouw adds,

> I think it's very dangerous for us to identify the will of God with a specific win. There may be all kinds of ways in which the outcome of a game could serve God's purpose, but God isn't a Michigan or Notre Dame fan. Football can be a way of serving God, and I think God cares about how people play the game. But I think we have to avoid identifying God with any partisan cause. (Nack 1998)

Those who critically respond to the possibility of God intervening in athletic activity express serious concerns about a God who could and would pick sides, manipulate bodily action on the field, and steer footballs into the right place. The issue put forth in this essay, however, is that these responses are already shielded before they are vented. These rejoinders and the attitude that animate them help reproduce a political theology that has, often unconsciously, disregarded the material body as a site of theological construction. The conviction held by many African Americans, and not by most whites, that God works through the bodies that have been treated as merely flesh and continue to be treated as such is ignored.

The double standard is, perhaps, most noticeable in the way in which Tim Tebow was treated during his rise in the 2011 season as quarterback of the Denver Broncos. Predictably, Tebow has been trained in the soft providential mode of thinking and speaking about God, especially as it relates to sports. What makes him unique in the world of the NFL that is stocked with obvious talent is that he has been widely regarded as bereft of the that same talent needed to be successful in the NFL. His lack of a job in the NFL heretofore has seemingly borne this out. Still, during the 2011 season, Tebow quarterbacked a team with midseason woes through last minute victories to a playoff win. His statistics during this run belied the win total—he was subpar statistically, but the Broncos still won games. This discrepancy prompted journalists across the spectrum to entertain the possibility that God was intervening in these games. Wary of fully committing to divine intervention in Tebow's achievements, some mainstream journalists still had no problem throwing around the word "miraculous" to describe what occurred in these games. His string of victories were "downright eerie" (Schefter 2012) for instance. Some were even led to question their atheism (Engber 2011). And Harvard researchers studied whether Tebow's wins were indeed miraculous based on statistical anomalies that dogged his history as a professional quarterback and these wins that followed. We can only assume that the word "miracle" was deployed in this study with a certain "tongue-in-cheek" attitude. But the mere fact that the term was used indicates the need for non-scientific terminology that may gesture towards the possibility of supernatural intervention in their supposed objective explanation (Bruce and Mooney 2013).

Miracles evince hard providentialism in unambiguous terms. But Tebow has consistently rejected any suggestion that God works in this way, despite the cacophony around him arguing otherwise. As he writes in his autobiography, "God gave us specific talents", "God receives joy when he sees me doing that with the skills he blessed me with", and that, "God has a plan for our lives" (Tebow and Whitaker 2011, p. 134). This is soft providential language to be sure, and Tebow is careful to walk up to the line that separates the hard and soft type, but he never steps over it.

> I know it sounds dumb to be praying over a football game ... I'm not sure God is into who wins or loses–He probably is more concerned with what you do in the process and what you will do with either result, to glorify Him and change the world by hopefully impacting one life. (Tebow and Whitaker 2011, p. 116)

To wit, after Tebow was credited by some for performing a miracle by touching and praying over a child who had a seizure in the stands, he demurred. "As far as me and miracles, no. But in the God that we serve, yeah, I do believe in miracles. I don't know what the situation was, but I know that the God that I get to serve is the God that always performs miracles in people's lives every day, all the time" (Peter 2016).

While Tebow's genuflecting on the field after a touchdown and giving praise to Jesus Christ immediately after games has been met with indignation by fans and the media alike wishing athletes remain silent on such matters, his theology has not faced the same judgment countenanced by black athletes for relying on providential language. Tebow's glorification language bleeds freely into his soft providential stance that God does intervene in football games by furnishing him with the strength to compete. Yet that providentialism is softened even further by subsuming into the glorification model. God's omnipotence and omnipresence are given credence out of deference, but a kind of theological guilt prevents Tebow and other athletes from taking their theology to its logical conclusion: if God prevents injuries on the field and strengthens you to win the game, why does God not also affect the outcome of the game, given what God supposedly cares about? By (perhaps artificially) insulating God from the need to care about the nitty-gritty of the game itself and who wins that game yet still maintaining that God does intervene in the game in a general sense, Tebow is able to walk down the established path that "correctly" conjoins God and sports.

He, like other white athletes, has no historical memory of having their body owned, depreciated, or feared. Therefore, his God has never been called on to act through his body; to redeem and purify it. God may bequeath Tebow disembodied gifts such as faith, talent, grit, strength, and a platform for evangelism but apparently never gives him first downs, touchdowns, or wins. His career and the reaction to it demonstrates the double standard applied to black and white athletes regards their religious expressions. God intervenes within the frameworks of both soft and hard providentialism, yet one type is deemed appropriate; the other is not. One involves God in all of life, sport included, though is unaware of the role that race plays in inoculating its speech from criticism; the other unabashedly involves God in minute bodily movements during a game and emerges out of the need to survive in a world and sport controlled by whites, yet is criticized nonetheless.

5. Conclusions

I have attempted to call attention to a racially based double standard as it has been applied to religious expressions of football players. Akin to the uneven treatment of trash-talkers in the NFL, hard providentialism that undergirds some religious talk by black football players is routinely ignored or deemed illegitimate while the soft providential language of white athletes and some black athletes gets a pass. A part of what makes this double standard so pernicious is that much of the discourse involving such talk is trafficked in the abstract that exists above the concrete historical and political forces that have given voice, and therefore communally justified hard providential theology within African American religious communities. Indeed, the authority required in order to contend that God does not care about football is granted only when the material sources of the speaker's theology are ignored. Because account is rarely taken by these cultured despisers of the historical and political factors that validate and absolutize their own critique (and allow a disregard of the sources of that which they critique) and protect it from the accusation that they use a double standard, their rules that regulate the relationship between God-talk and sport abide. Without entering into the debate of God's "proper "relationship to action on the field, this essay has ideally foregrounded the role that race plays in creating and sustaining such a double standard. Leaving it in the background or removing it from the frame altogether facilitates the deleterious perpetuation of the myth that we live in a post-racial society and that sport is the ultimate "colorblind meritocracy."

Funding: This research received no external funding.

Conflicts of Interest: The author declares no conflict of interest.

References

Blazer, Annie. 2015. *Playing for God: Evangelical Women and the Unitended Consequences of Sports Ministry*. New York: New York University Press.

Bruce, Chris, and Andrew Mooney. 2013. The Harvard Sports Analysis Collective. Available online: http://harvardsportsanalysis.org/2011/11/a-statistical-analysis-of-the-miracles-of-tim-tebow/ (accessed on 9 January 2019).

Buehring, Tom. 2019. I Think About Jesus Christ': These 4 NFL Players Are Keeping Faith at the Forefront of This Super Bowl. *CBN News*. February 3. Available online: https://www1.cbn.com/cbnnews/entertainment/2019/february/i-think-about-jesus-christ-these-4-nfl-players-are-keeping-faith-at-the-forefront-of-this-super-bowl (accessed on 3 March 2019).

Cox, Daniel, and Robert P. Jones. 2017. One-Quarter Say God Will Determine the Super Bowl's Winner—But Nearly Half Say God Rewards Devout Athletes. *PPRI*. January 30. Available online: https://www.prri.org/research/poll-super-bowl-women-sports-god-athletes-marijuana/ (accessed on 16 November 2018).

Curtis, Bryan. 2017. The Pretty Good Life of Tony Romo. *The Ringer*. March 8. Available online: https://www.theringer.com/2017/3/8/16041308/tony-romo-release-dallas-cowboys-dak-prescott-16e63faec1fb (accessed on 18 February 2019).

Du Bois, William Edward Burghardt, and Manning Marable. 1999. *The Souls of Black Folk*. New York: W. W. Norton and Company.

Engber, Daniel. 2011. Tim Tebow 2011: The Broncos Quarterback Is Making Me Question My Atheism. *Slate*. December 5. Available online: https://slate.com/culture/2011/12/tim-tebow-2011-the-broncos-quarterback-is-making-me-question-my-atheism.html (accessed on 18 February 2019).

French, David. 2015. The Ridiculous Movement to Take God Out of Football. *National Review*. September 22. Available online: https://www.nationalreview.com/2015/09/god-football-atheist-activists/ (accessed on 11 November 2018).

Hoffman, Shirl J. 1992. Evangelicalism and the Revitalization of Religious Ritual in Sport. In *Sport and Religion*. Edited by Shirl J. Hoffman. Champaign: Human Kinetics.

Howard, Johnette. 1996. Up from the Ashes Packer Reggie White Preaches that God Can Raise a Man to the Super Bowl and a Church from Ruins. *Sports Illustrated*. September 2. Available online: https://www.si.com/vault/1996/09/02/217042/up-from-the-ashes-packer-reggie-white-preaches-that-god-can-raise-a-man-to-the-super-bowl-and-a-church-from-ruins (accessed on 17 October 2018).

Klemko, Robert. 2017. Available online: https://www.si.com/mmqb/2017/06/15/nfl-team-survey-player-upbringings-race-class-2016-election (accessed on 8 May 2019).

Krattenmaker, Tom. 2010. *Onward Christian Athletes: Turning Ballparks into Pulpits and Players into Preachers*. Lantham: Rowman & Littlefield.

Leonard, David J. 2017. *Playing While White: Priviledge and Power on and off the Field*. Seattle: University of Washington Press.

Nack, William. 1998. Does God Care Who Wins the Super Bowl? *VAULT*. January 26. Available online: https://www.si.com/vault/1998/01/26/238058/does-god-care-who-wins-the-super-bowl-many-packers-and-broncos-think-the-lord-will-decide-the-outcome-theologians-beg-to-differ (accessed on 14 November 2018).

Parke, Caleb. 2019. Russell Wilson Gives Glory to God after Becoming NFL's Highest-Paid Player. *Fox News*. April 16. Available online: https://www.foxnews.com/faith-values/russell-wilson-gives-glory-to-god-after-becoming-nfls-highest-paid-player (accessed on 22 June 2019).

Peter, Josh. 2016. Tim Tebow on Miracles, Helping Arizona Fall League Fan in Need. *USA Today*. October 13. Available online: https://www.usatoday.com/story/sports/mlb/mets/2016/10/12/tim-tebow-on-helping-arizona-fall-league-fan/91980992/ (accessed on 27 October 2018).

Pinn, Anthony B. 2003. DuBois' Souls: Thoughts on "Veiled Bodies and the Study of Black Religion". *The North Star: A Journal of African American Religious History* 6: 1–5.

Raboteau, Albert J. 1980. *Slave Religion: The "Invisible Instition" in the Antebellum South*. New York: Oxford University Press.

Schefter, Adam. 2012. Available online: https://www.espn.com/nfl/story/_/page/10spot-divisional/tim-tebow-phenomenon-gets-eerie--adam-schefter-10-spot (accessed on 16 October 2018).

Schneck, Tim. 2013. The Theology of Ray Lewis. *Huffington Post*. February 2. Available online: https://www.huffpost.com/entry/the-theology-of-ray-lewis_b_2617107 (accessed on 14 January 2019).

Schuster, Shawn. 2014. Former NFL Star Kurt Warner Catches Criticism for His Christian Beliefs, Told He Should be 'Lined up and Shot'. *The Gospel Herald*. November 10. Available online: https://www.gospelherald.com/articles/53130/20141110/former-nfl-star-kurt-warner-catches-criticism-for-his-beliefs-told-he-should-be-lined-up-and-shot.htm (accessed on 22 March 2019).

Tebow, Tim, and Nathan Whitaker. 2011. *Tim Tebow: Through My Eyes*. New York: HarperCollins.

Thompson, Chris. 2018. Ray LewisThe Theology of Ray Lewiss Hall Of Fame Speech Was Actual Frothing Madness. *Deadspin*. August 8. Available online: https://deadspin.com/ray-lewiss-hall-of-fame-speech-was-actual-frothing-madn-1828116271 (accessed on 16 November 2018).

Van Bibber, Ryan. 2018. What the hell is Ray Lewis Talking About? *SBNation*. August 5. Available online: https://www.sbnation.com/nfl/2018/8/4/17630430/ray-lewis-hall-of-fame-speech-ravens-what-is-he-talking-about (accessed on 14 November 2018).

Religions **2019**, *10*, 470

Zirin, Dave. 2015. The Age Gap in Religion around the World. *The Nation*. September 21. Available online: https://www.thenation.com/article/getting-god-out-of-football/ (accessed on 3 November 2018).

The Age Gap in Religion around the World. 2018. Available online: https://www.pewforum.org/2018/06/13/the-age-gap-in-religion-around-the-world/ (accessed on 4 April 2019).

religions

MDPI

Article

Affect Theory, Religion, and Sport

Eric Bain-Selbo

Department of History, Political Science, and Philosophy, Indiana University Kokomo, Kokomo, IN 47405, USA; ebainsel@iu.edu

Received: 26 June 2019; Accepted: 10 July 2019; Published: 31 July 2019

check for updates

Abstract: Affect theory has made important contributions recently to the study of religion, particularly drawing our attention away from ideas and practices to the emotional or affectual experience of religion. However, there is a danger that affect theory may become yet another "protective strategy" (to use a term from philosopher of religion Wayne Proudfoot) in academic wars about the nature of religion. As a consequence, there is a danger that affect theory will become too restrictive in its scope, limiting our ability to use it effectively in investigating "religious" or "spiritual" affects in otherwise secular practices and institutions (such as sport). If we can avoid turning affect theory into a protective strategy, it can become a useful tool to provide insights into the "spirituality" of sport.

Keywords: affect theory; sport; religion; spirituality; phenomenology of religion

---

This essay is speculative in nature. Relatively speaking, affect theory is still new in terms of its application to the study of religion, and its full benefits may not come to fruition for some time. However, it holds great promise as a tool in our efforts to understand the emotional or affectual experience of religion as opposed to its mere doctrines and practices. In addition, it holds great promise as a tool for understanding the religious or spiritual dimensions of secular or seemingly non-religious cultural phenomena—in particular, in our case, sport. However, the utility of affect theory can only be realized fully if we avoid turning it into what may be called a "protective strategy".

In this essay, I will argue that affect theory can be an effective tool in the study of religion and popular culture (including sport), but only if we understand and use it in a particular way. After summarizing a common understanding of affect theory, I turn to philosopher of religion Wayne Proudfoot and his powerful critique of "protective strategies" in the study of religion. In the process, I will show how Proudfoot's critique can help us in maintaining a proper view of affect theory and its application to the study of religion. I conclude by noting how an appropriate view of affect theory is critical to the study of the religious or spiritual dimensions of sport.

1. Affect Theory

Affect theorists often describe their object of study in negative terms—what affects are *not* or what they are *not* connected to. For example, two primary characteristics of affects are, first, that they are pre-discursive or *not* directly related to or dependent upon language and, second and relatedly, that affects are *not* to be understood as necessarily inferior to reason.

Donovan O. Schaefer describes affects as "the flow of forces through bodies outside of, prior to, or underneath language".[1] In other words, "affect or affects can be understood as the propulsive elements

---

[1] (Schaefer 2015).

of experience, thought, sensation, feeling, and action that are not necessarily captured or capturable by language or self-sovereign 'consciousness'".[2] Take the example of joy. John Corrigan writes:

> So, affect theorists see in the smile a sign of an affective *fact*, the affect of joy. That joy, displayed on the face, is not something that persons have to talk themselves into. It is a physically embodied emotion, but not one that requires the discourse of culture—however those are defined and displayed—in order to take place.[3]

The idea that emotions or affects must be dependent on language is part of what Schaefer calls the "linguistic fallacy". He explains:

> The linguistic fallacy affirms that depth can't exist without language—that we can't want things without being told that we want them, without deciding that we want them, or without learning to want them. This is the presupposition of classical psychological behaviorism as much as textualism. But affect theory suggests that our animal intimacy with the world precedes constitution inside a linguistic frame—that there are "Proustian nooks" that pull us into the world without the application of language. The relationship between the affect and power moves bodies transversally through and across the grids of language, consciousness, or cognition. The compulsions of affect are better understood as addictions, as thick passions for bodies, objects, and relationships.[4]

Not only are affects independent of language, but they often are more powerful. "Affects are not passive receptors of inconsequential feeling that serve as window dressing on the linguistic architecture of power", Schaefer concludes, "power moves bodies through the pulsing of mobile, uneven affective systems. The linguistic *I* is a figurehead monarch in this field of recalcitrant attachments."[5]

Because language and reason are so bound together, it is not surprising that affects are seen as being in a complicated, if not contentious, relationship to reason. "To study affects is to explore nonsovereign bodies, animal bodies", Schaefer argues, "bodies that are propelled skittering forward by a lattice of forces rather than directed by a rational homunculus".[6] He adds:

> Affects invert the metaphysical emphasis on the human's rational sovereignty over its body, retracing us as nests of animal becoming, finding pleasure in spinning out of control. Affective economies are directed by compulsions—by autotelic forces that derail the abacus of rational self-interest.[7]

However, it is not the case that we now have a dichotomy with elements or aspects (affects and reason) that never interact and remain solely in their domains. In other words, there is not feeling and thinking separately. There are feelings to thinking and feelings to knowledge production, as well as the holding of knowledge. "Within the framework of affect theory, affect animates every aspect of embodied life, including the ostensibly affect-neutral domain of knowledge-production", Schaefer insists. "Affect theory prompts us to ask not just what we know, not just how we know, but *how knowledge feels*".[8]

By elevating the role of affects vis-à-vis reason and knowledge, affect theorists deconstruct a powerful and persistent view of the hierarchical structure of the human being. This hierarchy certainly dates back millennia and posits our rational nature as superior and controlling and our affectual or

---

[2] Schaefer, *Religious*, p. 23.
[3] (Corrigan 2017).
[4] Schaefer, *Religious*, p. 15.
[5] Schaefer, *Religious*, p. 94.
[6] Schaefer, *Religious*, p. 24.
[7] Schaefer, *Religious*, p. 166.
[8] (Schaefer 2017).

emotional nature as inferior and in need of control. "Separating reason and emotion denigrates the embodied nature of cognition, resorts to an ancient dualism of mind and body, and erects a hierarchy of thought, feeling, and body that skews the explanation of human behavior as properly rational", David Morgan observes. "The resulting dualism is strongly disposed to regard emotion as suspect for its inherent tendency to move one independent of reason".[9]

Liberated from language and rationality, affects can be seen as "ends in themselves".[10] What Schaefer means by this is that affects cannot be reduced to other phenomena. For example, they are not merely the epiphenomena of language, culture, or rationality. In this sense, we might see an analogy between the way that affect theorists treat affects as *sui generis* and the way that Mircea Eliade, perhaps the 20th century's most prominent scholar of comparative religion, treats the sacred as *sui generis*.[11] As we will see below, there are two related questions: Are affects *sui generis*? Or, are *religious* affects *sui generis* (though other affects may not be)? Answering affirmatively to the second question in particular might provide the theoretical basis for a "protective strategy" for religious phenomena that isolates or protects certain claims about religious affects from rational inquiry.

The independence of affects from language and reason is critical when studying religion. Affect theory draws us to these powerful elements in human beings—elements that are not simply activated by or determined by language and rationality but exist independently and prior to or outside of all language and reason. Schaefer insists that "affects are not simply to be understood as passive channels activated by the play of language hovering over them. Rather, affects surge through bodies, producing semistable structures that become the tough, raw materials of religion".[12]

Part of the problem with the past study of religion has been its focus on language—the "linguistic fallacy". "The linguistic fallacy misunderstands religion as merely a byproduct of language", Schaefer writes, "and misses the economies of affect—economies of pleasure, economies of rage and wonder, economies of sensation, of shame and dignity, of joy and sorrow, of community and hatred—that are the animal substance of religion and other forms of power".[13] Looking behind or before language leads scholars then to a new and very different approach to the study of human beings and culture in general and religion in particular. For Schaefer, such an approach even closes the gap between human beings and other animals. He insists:

> [A]t the emotional and preemotional levels, affects are the flexible architecture of our animal lifeways, the experiential shapes that herd together and carry religion on their backs. Affect theory makes available a set of approaches to religion that work through animality by probing the thick forms moving outside of the narrow lighted circle of language.[14]

Affects are like the foundation of a house upon which reason, desires, and human action build cultural phenomena—for example, religion.

For Schaefer, affects are both the prelinguistic foundation of religion and a consequence of religious activity. "Religion as a composite of compulsions is made possible by existing, intransigent bodily technologies", he writes, "but it also motivates, activates, and drives those technologies".[15] Rather

---

9 (Morgan 2017).
10 Schaefer, "Beautiful", p. 79.
11 Schaefer also is interested in how affects are related to power. For him, "Affects fuse together to shape the planes of interface between bodies and power" (Schaefer, "Beautiful", p. 77). He also states, "Affect theory in all its forms is designed to profile the operations of power outside of language and the autonomous, reasoning human subject" (Schaefer, *Religious*, p. 23). Schaefer's work in this direction is interesting and worth engaging but not critical for our purposes.
12 Schaefer, *Religious*, p. 39.
13 Schaefer, *Religious*, pp. 9–10.
14 Schaefer, *Religious*, p. 24.
15 Schaefer, *Religious*, p. 134. He adds: "Religion is an extraordinarily powerful distribution network through the global nervous system of affect" (Schaefer, *Religious*, p. 175).

than arguing for a simplistic cause-and-effect position (religion causes affects or affects cause religion), Schaefer is defending a more circular understanding of the relationship.[16]

In situating affect theory in the broader history of the study of religion, it is reasonable to wonder if it is a new version of the phenomenological approaches of figures such as Rudolf Otto or Mircea Eliade. In one sense, affect theory can embrace that phenomenological history. Similar to Otto's "idea of the holy" or Eliade's experience of the sacred, affects have a *sui generis* character. On the other hand, someone such as Schaefer is hesitant to embrace the "ahistorical metaphysical essentialism of Eliade"[17]—insisting on the cultural context and historical grounding of affects. The question is the following: Are there *religious* affects or simply affects that tend to occur as a consequence of religious practices or in religious settings? If the latter, then we certainly have something very different than the phenomenological approach of Otto or Eliade. However, if it is the former, affect theory would seem to come particularly close to the kind of *sui generis* arguments of scholars such as Otto and Eliade. Before answering the question, however, it will be helpful to look at the work of Wayne Proudfoot, a philosopher of religion who offers a powerful critique of the phenomenological approach—one that might serve as a warning to affect theorists.

2. Wayne Proudfoot on Religious Experience and Protective Strategies

Affect theory is far from the first attempt to draw our attention away from language and reason and focus on emotion or feeling instead. In the 19th century, theologians and scholars of religion started to identify feeling or emotion as that which was most characteristic of religion. Wayne Proudfoot argues that this primarily occurred for two reasons. First, it was deemed that feelings or emotions are more grounded in the lived experience of adherents than is doctrine. Feelings or emotions are more powerful than dogma. Second, the move to feelings or emotions helped to avoid a rationalist critique of religion. At least since the Enlightenment, religious doctrine had been subject to powerful philosophical criticism. Indeed, in the 19th century, through the 20th century, and continuing into the 21st century, many intellectuals predicted and still predict the demise of religion as individuals and whole societies become more rational. However, reason often reaches a certain limitation (so it is thought) when confronted with feelings or emotions. At the very least, feelings or emotions do not seem subject to the same level of rational critique as does doctrine.[18] Our feelings or emotions are natural or intrinsic elements of the human condition in the same way that some theorists speak of affects today.

Christian theologians such as Friedrich Schleiermacher and Rudolf Otto, whose interests included the more general study of religion as well, were instrumental in moving the study of religion in this theoretical direction. The former emphasizes the role of non-cognitive feelings or intuitions as foundational to religion, while the latter describes feelings of awe, energy, and mystery (among others) that are constitutive of religious experience. William James, the early 20th-century philosopher and psychologist, also emphasizes the subjective experience of religious phenomena as central to the study of religion. His *Varieties of Religious Experience* is a seminal work in the typology of religious experience. He describes religion not in terms of doctrines or institutions but as *"the feelings, acts, and experiences of individual men in their solitude, so far as they apprehend themselves to stand in relation to whatever they may consider the divine"*.[19] Eliade, of course, also focuses on subjective experience. In his phenomenology of religion, Eliade argues for the qualitatively greater experience associated with the sacred as opposed to what is associated with the profane.

---

[16] Religion also is an area where Schaefer analyzes power. He notes, "Religion, like other forms of power, moves bodies by creating affective ligatures between bodies and their worlds" (Schaefer, *Religious*, p. 179). In other words, religions "like other formations of power, are reservoirs of compulsory links to the world" (Schaefer, *Religious*, p. 207). In the end, he concludes: "Affects wordlessly crawl through our bodies on the way to the world, producing systems of power as they do—some of which get called religious" (Schaefer, *Religious*, p. 218).

[17] Schaefer, *Religious*, p. 8; also see Schaefer, *Religious*, pp. 54, 59.

[18] (Proudfoot 1985).

[19] (James 1961).

Wayne Proudfoot offers a compelling critique of this historical and theoretical development in the study of religion. He finds that many of the approaches to understanding religious experience simply protect that experience against rational investigation and criticism. One cannot rationally dismiss, for example, that which precedes or is beyond reason (a popular argumentative strategy, of course, for those who defend claims about the existence of God or gods). However, these "protective strategies" fail philosophically. Take, for example, Proudfoot's critique of Schleiermacher—one that can be (and is) extended to other theorists as well.

Schleiermacher describes religious experience as an immediate apprehension of the divine or religious object or being. By making the religious experience immediate, Schleiermacher preserves it against the argument that it is the "idea" or "thought" of the divine that causes the experience and, extending this, that the divine is *merely* an "idea" or "thought" and thus has no external reality (or, at least, following Kant, that we cannot know that reality). Religious consciousness, according to Proudfoot's reading of Schleiermacher, "is both intentional, in that it is directed toward the infinite as its object, and immediate. It is not dependent on concepts or beliefs, yet it can be specified only by reference to the concept of the whole or the infinite".[20] Taken together, however, these claims are contradictory. "If the feeling is intentional", Proudfoot writes, "it cannot be specified apart from reference to its object and thus it cannot be independent of thought".[21] In other words, Schleiermacher "defends the incoherent thesis that the religious consciousness is both independent of thought and can only be identified by reference to concepts and beliefs".[22] He cannot have it both ways. Either religious experience is truly immediate (i.e., not mediated by thought), in which case it becomes hard to identify it vis-à-vis an intentional object (thus, why even call it "religious"), or it is mediated by thought, in which case it is not immediate and thus open to rational inquiry and criticism.

There are at least two key points that come out of Proudfoot's critique of Schleiermacher—again, points that can be extended to subsequent theorists who attempt (either explicitly or implicitly) to protect religious experience from critical inquiry. First, Proudfoot makes the point that religious language is both expressive and formative of experience. Religious language may describe an experience but the words and ideas that we associate with religious language also cause or at least shape the experience itself. Thus, Proudfoot argues that religious language "is not only the expressive, receptive medium Schleiermacher takes it to be. It also plays a very active and formative role in religious experience".[23] The second (related) point is that it is illegitimate to separate religious feeling or emotion (in short, religious experience) from thought. Proudfoot admits that Schleiermacher "is correct to view primary religious language as the expression of a deeply entrenched moment of consciousness", but he is "incorrect to portray that moment as independent of thought and belief. Schleiermacher has mistaken a felt sense of immediacy for a guarantee that piety is not formed or shaped by thought or inference".[24] Do affect theorists make a similar mistake?

The question of the cognitive status of feelings or emotions is critical and deserves more of our attention here. Take the example of anger. Anger is not an immediate emotional experience, unmediated by thought. How do we know when someone is angry? It certainly is not because we empathically feel what they are feeling. We ascribe anger to them based on the public evidence, based on how they are reacting. This ascription will include our interpretation of their actions relative to our understanding of the entire context in which they are acting. Frank seems agitated, and I interpret that as anger, because I know that he has just been told by his boss that he will not be receiving a raise this year. In short, my ascription of anger to someone is a consequence of my reflection about all sorts of

---

20 Proudfoot, p. 11.
21 Proudfoot, p. 11.
22 Proudfoot, p. 18.
23 Proudfoot, p. 40.
24 Proudfoot, p. 36.

pieces of evidence provided to me. What is even more interesting is that our self-ascription of anger is very similar. As Proudfoot argues:

> I don't appeal to private inner states in ascribing emotions to myself any more than I do in ascribing them to others. I often come to know what I am feeling by interpreting physiological changes or my behavior in exactly the same way in which another might interpret them if the data were available to him.[25]

Coming out of a meeting with my boss, I may notice that my heart is racing, and my teeth are clenched. These physiological changes alone are not enough for me to determine that I am angry. Certainly, they can contribute to such a determination, but I will come to ascribe anger to myself by interpreting the situation, realizing that my boss is a jerk and that I have been treated poorly or unfairly by her. Indeed, I may not even realize that I am angry based just on increased heart rate and clenched teeth. I may not even notice these physiological changes. To that extent, the physiological changes alone do not constitute anger. Only when interpreted and understood within a context of perceived sleight, injury, etc. can we connect these physiological changes to the feeling or emotion of anger. The word "anger" is not merely a description of our experience, but is an interpretation of it. Proudfoot concludes that emotion words "are employed, not as simple descriptions of bodily changes, behavior, or dispositions to behave, but as interpretations and explanations of those phenomena".[26] Even more, the interpretation is constitutive of the very experience.

Proudfoot turns to the psychological research of Stanley Schachter to further his argument. Schachter's experiments confirm that physiological changes alone are not clear indicators of particular emotions or feelings. In other words, the same physiological changes may be interpreted in different ways depending on the person who is experiencing them and the context in which they occur. However, what is the relevance of Schachter's work for Proudfoot's interest in religious experience and for our own concerns here? "Given the results of Schachter's experiments," Proudfoot concludes, "it seems quite plausible that at least some religious experiences are due to physiological changes for which the subject adopts a religious explanation".[27] Thus, if understood in the same way that we should understand other emotions or feelings, the physiological changes (the felt experience) of the religious experience are in fact religious to the extent that they are interpreted religiously—they are not intrinsically religious.

This understanding of what constitutes an experience is critical to Proudfoot's distinction between descriptive and explanatory reduction in the study of religious experience. Descriptive reduction is "the failure to identify an emotion, practice, or experience under the description by which the subject identifies it. This is indeed unacceptable".[28] Proudfoot uses the example of a hiker seeing a bear in the woods, the sighting leading to an experience of fear in the hiker. As it turns out, it really was not a bear but a tree stump instead. It would be a case of descriptive reduction to claim that the hiker was in fear of a tree stump. The tree stump indeed was the object he saw, but he thought it was a bear. To say that he was afraid of a tree stump would be to fail to make sense of the story. Still, it would not be wrong to say that the cause of his fear was a tree stump that looked to him (perhaps at a distance, through some fog or mist, etc.) like a bear. We also must take account of the hiker's understanding of the danger of bears or even his past experience with bears (maybe he was mauled as a child?). Including this information in our account of the incident would be leading us toward explanatory reduction, "offering an explanation of an experience in terms that are not those of the subject and that might not meet with his approval. This is perfectly justifiable and is, in fact, normal procedure".[29] It might be

---

[25] Proudfoot, pp. 92–93.
[26] Proudfoot, p. 93.
[27] Proudfoot, p. 102.
[28] Proudfoot, p. 196.
[29] Proudfoot, p. 197.

that the hiker refuses to believe that there was no bear—that what he really saw was a tree stump. I certainly am not obligated simply to accept his account of the experience, especially if I have strong evidence supporting the claim that what he saw really was a tree stump. In other words, I have to take him seriously when he says that he saw a bear and that this is what made him afraid, but I need not accept that as the final explanation of the event. I clearly have solid evidence of the tree stump, and I might even be able to make the case that the hiker suffers from paranoia vis-à-vis bears (thus leading to the mistaken identification of the tree stump as a bear). Proudfoot concludes:

> Where it is the subject's experience which is the object of study, that experience must be identified under a description that can plausibly be attributed to him.... The explanation the analyst offers of that same experience is another matter altogether. It need not be couched in terms familiar or acceptable to the subject. It must be an explanation of the experience as identified under the subject's description, but the subject's approval of the explanation is not required.[30]

All of this can be applied to religious experience in the following way. If a religious adherent claims to have had a religious experience of God's love—a uniquely *religious* love unlike any other experience of love—then any investigation of this experience must begin with the adherent's description. However, a full explanation of the experience may entail an account of the full context in which the experience occurred in order to more accurately identify the factors or causes that gave rise to the experience. For example, perhaps it was an emotionally taxing period in the adherent's life. Maybe she was part of a prayer group that emphasized the experience of God's love during its communal activity. Maybe she was on medication that fostered such loving emotions. In short, many factors may have led to her experience of God's love, and we need not accept the *sui generis* character of that experience any more than we need accept the reality of God. Naturally, she might reject these latter explanations. However, while we must begin with her description, we need not end up there. As Proudfoot argues, "To require that any explanation of a religious experience be one that would be endorsed by the subject is to block inquiry into the character of that experience".[31] In other words, if all I can do is to accept her description of the experience, then there is no room for any other explanation nor even for an investigation of her experience in the first place.

Proudfoot concludes that the "distinguishing mark of a religious experience is not the subject matter but the kind of explanation the subject believes is appropriate".[32] Another way of putting this is that it is not the content (increased heart rate, feelings of elation, forms of ecstasy, etc.) of the experience that defines it, but the explanation we give to that content. For example, in the case of experiences surrounding the participation (either as an athlete or spectator) in sporting events, it very well could be that the participants have similar physiological and psychological experiences that religious practitioners have, but the former are not having "religious" or "spiritual" experiences, because they simply do not label them that way as do the latter. If, for example, the participants in the sporting event had a different understanding of what religion is or what a spiritual experience is, perhaps they more likely would apply these terms to describe their experiences and, thus, those experiences legitimately could be considered religious or spiritual.

The approach Proudfoot represents certainly opens up popular culture generally and the experience of the sporting event more specifically to a deeper and more thorough investigation. There are other theorists whose work supports or is supported by such an approach. Mihaly Csikszentmihalyi champions the psychological concept of "flow". Flow involves the immersion of the individual psyche in an activity that is productive, creative, and personally valuable. Flow is "the state in which people are so involved in an activity that nothing else seems to matter; the experience itself is so enjoyable

---

[30] Proudfoot, p. 195.
[31] Proudfoot, p. 200.
[32] Proudfoot, p. 231.

that people will do it even at great cost, for the sheer sake of doing it".[33] Flow experiences can occur in all sorts of activities, ranging from painting a picture to dancing to making a cabinet. They can occur in religious settings. They can also occur in sports. "Play, art, pageantry, ritual, and sports are some examples [of flow]", Csikszentmihalyi writes. "Because of the way they are constructed, they help participants and spectators achieve an ordered state of mind that is highly enjoyable".[34] He adds that in the sporting event "players and spectators cease to act in terms of common sense, and concentrate instead on the peculiar reality of the game".[35] Thus, the flow experience is common or universal whether it occurs in the art studio, the church, or the stadium—it is just that the contexts will shape how we label the flow experience.

If the content of the experience is similar (ecstasy, "flow", etc.) but we simply label it differently, then it should not surprise us when people decide that it is appropriate and perhaps necessary to use religious language to more accurately describe the experience of the sporting event. They have come to believe that the content of the experience is similar to if not identical with those experiences described by religious practitioners (for example, mystics). Howard Slusher insists: "Something of faith, something of peace, a touch of power, a feeling of right, a sense of the precarious—all of these and more is what *real spirit* of sport *is*".[36] He acknowledges the mystical dimension of sport and religion, concluding that both "open man towards the acceptance and actualization of being".[37]

It is important to remember, however, that Proudfoot makes a convincing case that our labels of experiences do not merely describe them but help to constitute them. There still might be something different about the religious experience—because it in part is constituted by religious concepts and ideas—that separates it from the often equally powerful experiences at the sporting event. However, here we are pushed to our reflective limits and the recognition that we are now probing psychological and existential areas where we cannot have anything close to definitive answers. Is there a difference between the ecstasy in the pews and the ecstasy in the bleachers? Given the complexities and subtleties of psychological states, we probably can never know. However, we should not begin our investigations with theories that are "protective strategies", that separate out some experiences as *sui generis*, that close off avenues of inquiry about the similarities or even identical natures of human experiences.

3. Critique of Affect Theory

So, what of affect theory? Does Proudfoot's critique of the phenomenological approach pose a threat to affect theory as an approach to the study of religion? As with a lot of questions, it depends. It depends most significantly on whether we think there are fundamental affects as part of the human condition or that there are specific affects that are irreducibly religious.

If the claim by affect theorists is that there are affects that are exclusively religious affects, then we return to something like the *sui generis* argument of Eliade and others—the kind of argument that Proudfoot identifies as a "protective strategy". On the other hand, if we are talking about a set of general affects, some of which encourage and/or are activated by religious thought and actions, then we recognize the possibility that that subset of affects may also encourage and/or be activated by thought and actions that may be stereotypically secular in character.

Schaefer is not always clear on this issue. Generally, he writes about affects as universals that can be connected to a wide variety of practices, institutions, and structures. At other times, however, he suggests the existence of specifically *religious* affects. Indeed, even the frequent reference to *religious* affects (including in the title of his book) suggests that there are actual affects that are religious in nature. Beyond just the use of the adjective "religious", however, there are passages that seem to

---

33 (Csikszentmihalyi 1990).
34 Csikszentmihalyi, p. 72.
35 Csikszentmihalyi, p. 72.
36 (Slusher 1993).
37 Slusher, p. 191.

claim the existence of specific religious affects. For example, Schaefer states that affects "surge through bodies and compose themselves in religious forms".[38] What are these "religious forms"? The phrase leads one to believe that affects change somehow into forms that are specifically religious. But how? Is simple joy different than religious joy? In another passage, Schaefer goes further, making it sound like there are permanent or eternal religious affects. In *Religious Affects*, he writes of "the multitude of subterranean ways that religion flows through our bodies".[39] Is religion a "thing" that flows through bodies? We would hardly think so. Does Schaefer mean then that religious affects flow through bodies? In short, both passages suggest a return to the sort of *sui generis* arguments of the older phenomenological approach of Eliade and others.

Schaefer, in fact, tries to separate himself from this phenomenological approach. He imagines a line that goes roughly from Eliade and Otto to J.Z. Smith to affect theory. He considers Smith to be an important advance beyond the phenomenological tradition represented by Eliade and Otto and sees affect theory as an important advance beyond Smith.

What would Proudfoot say about the affect theory we find in Schaefer? Clearly, the claim that there are pre-linguistic affects that are "religious" (whatever that means) leads us right into the same problems that Proudfoot critiques in the phenomenological tradition. If they are pre-linguistic, then why call them "religious". In this sense, *religious* affects are akin to Eliade's experience of the sacred or Otto's *mysterium tremendum*. However, if affects are viewed as merely emotional experiences or states that can be elicited or activated in a number of different settings—some stereotypically religious and others seemingly secular—then we open up a much richer and defensible approach to the study of culture and cultural institutions and the human experience of them. In this regard, I agree with Abby Kluchin who warns, "Affect theory cannot afford to be so enchanted with itself—with its own bewitching linguistic formulations to capture the nonlinguistic—that it unwittingly re-enchants our thoroughly disenchanted world".[40]

In his essay on religion, sport, and nationalism, David Morgan attempts to "thread the needle" on the problems raised above. "My contention will *not* be that sport or national piety is the new religion of modern society," he writes, "but rather that there is no need for the social or cultural analyst to erect a strong distinction among the three. They are not fully discrete any more than they are merely interchangeable".[41] In other words, the affects associated with these three areas of cultural life are neither unique and irreducible nor exactly the same and interchangeable. There may not be a discrete set of religious affects wholly different from all other affects, but we still can talk about affects that either arise in a religious context or have a religious "flavor" to them.

Morgan relies upon the notion of "sacralization". "*Sacralization* is a procedure at work in any number of cultural activities, including but not limited to religion", he notes. "Thus, when fans say *soccer is my religion*, they may be understood to say something like *soccer is how my people and I feel our common identity*. They might say the very same of their nation and their religion".[42] Morgan then rejects the notion that any of these categories needs to be "reduced" to the other—that sport is *really* religion or that religion is *really* nationalism or even that nationalism is *really* sport. He insists that "sacralization is not synonymous with religion nor is it essentially religious, but is rather the pervasive social mechanism for making something, someone, or someplace special. It happens in sports, art, politics, commerce, the family, and religions".[43] In a sense, the point is that the noun (sport, art, "the sacred", etc.) is less important than the verb (sacralize)—than the process by which something is made special. Morgan concludes that rather than reducing other cultural phenomena to

---

[38] Schaefer, *Religious*, p. 117.
[39] Schaefer, *Religious*, p. 209.
[40] (Kluchin 2017).
[41] Morgan, p. 223.
[42] Morgan, p. 228.
[43] Morgan, p. 230.

the phenomenon of religion or vice versa, "we get further by describing such modern experiences as sport, nationalism, and religion as ritual practices that generate powerful cultures of thought and feeling that enable moderns to imagine the bonds of affection that tell them who their group is and what matters to them".[44]

Morgan cites Ann Taves as a kindred spirit in this regard, and certainly her work makes a substantial contribution. She argues against the *sui generis* approach typical of phenomenologists such as Eliade—the idea that sacredness or holiness is some irreducible reality in the world. She argues that the "basic problem with the sui generis model is that it obscures something that scholars of religion should be studying: that is, the process whereby people constitute things as religious or not".[45] In order to get away from problematic terms such as sacred and holy, Taves instead writes of "specialness". Things, people, places, actions, or times that are special can vary greatly. Many of these are stereotypically religious (e.g., a temple, a priest, a ritual, etc.), but what is special can also appear in contexts that are secular. In other words, the process of making something special is the big umbrella under which religion is but one institution—along with music, sport, nationalism, and much more. Special things, for Taves, are the "building blocks" of religion but also of many other cultural phenomena.[46]

Both Taves and Morgan, then, suggest a world of experiences or affects that are more universal in character and fluid between cultural phenomena. They represent a very different position than those who might posit the existence of uniquely *religious* experiences or affects.

4. Conclusions

It might help for us to imagine a spectrum of positions to take on the status of religious affects. At one extreme is the claim that there are affects that *are* religious—not reducible to other affects, unique, *sui generis*, etc. We might imagine Eliade and Otto occupying this position (if they were alive to join in the discussion), and Schaefer sometimes seems to drift in this direction (though, as stated, his position is not always clear). Thus, one might argue that there is a distinct religious experience of transcendence, of feeling part of something larger than oneself, and that this experience is unlike any similar sorts of experiences in different contexts.

At the other extreme is the claim that "affects *are* affects"—that affects are basic building blocks (to borrow Taves' term) of the human condition that can be found in religious contexts but in many other contexts as well. In this sense, *religious* affects are not unique affects. Rather, the word "religious" simply designates where the affects arise or are manifested or how they are interpreted rather than any irreducible or *sui generis* character that they have. Thus, one would argue that the experience of transcendence, of feeling part of something larger than oneself, is an experience that can happen in a religious context, but it may also happen at a political rally or in a sports stadium. The affect is the same, though the context might change. This position is supported by the work of Proudfoot, and Taves certainly can be read to lean in this direction.

Between these two extremes, of course, are myriad other positions—Morgan being but one example. However, it is hard to occupy a middle position on this issue, because the two extremes require an either/or choice. Either religious affects are *sui generis*, or they are not. Proudfoot opens the door to such middle positions by noting how language and/or context might shape experience rather than serve as mere interpretation. However, as I noted earlier, we are then "in over our heads" in terms of really figuring out and disentangling cause and effect here. We are beyond being able to come up with definitive answers. That said, and while I cannot fully defend the claim here (the claim may even

---

[44] Morgan, p. 238.
[45] (Taves 2009). It should be noted that Taves recognizes a fair amount of agreement between her position and Proudfoot's work (Taves 2009, p. 93).
[46] Taves, p. 162.

be indefensible or at least unproveable), I believe Morgan's position ends up being a theoretically untenable compromise.

In the end, I support Proudfoot's side of the spectrum. There certainly is a danger that affect theory turns us back to the problematic kind of protective strategy that Proudfoot powerfully identifies in the phenomenological tradition. I argue that the study of the full range of human affects in popular culture is only possible by avoiding that protective strategy and that the study of the full range of human affects in popular culture also helps us to avoid that protective strategy. Seeing the experiences of joy, "flow", and transcendence in a sporting event "disenchants" those experiences—frees them from the limitations of specifically religious contexts. If there is such a thing as *religious* joy, for example, that can mean one of two things. First, that there is an intrinsically religious character in a particular kind of joy that makes it different from any other joy-like affect. Here, we are in the neighborhood of Eliade and other classical phenomenologists. In that neighborhood, our questioning or inquiry into that joy is limited, for we first must accept the intrinsically religious character and that character, at least, cannot be questioned. Second, one might say that *religious* joy is the joy that occurs in a religious context. However, then the very phrasing "religious joy" is misleading, for what we really mean is that one might experience joy as part of a religious ritual or being in a religious building, but one also could experience that same joy in other contexts. Thus, instead of calling it *religious* joy, we simply should talk about joy as it may occur in a religious context. Now, we are in the neighborhood of Proudfoot—the neighborhood that I prefer. In this neighborhood, we can query more fully the nature of this joy—how it arises and how its appearance in a variety of contexts (religious or secular, in the temple or the stadium) can lead to powerful insights about it.

As a key element of popular culture, the study of sport by religious studies scholars is most fruitful when we avoid protective strategies (either ones we inherit or new ones of our own creation) and that same study implicitly makes the case against those protective strategies. Affect theory can be a powerful tool for religious studies scholars studying popular culture more generally and sports more specifically, but only if it avoids becoming a new protective strategy.

Funding: This research received no external funding.

Conflicts of Interest: The author declares no conflict of interest.

References

Corrigan, John. 2017. Introduction: How Do We Study Religion and Emotion? In *Feeling Religion*. Edited by John Corrigan. Durham: Duke University Press, p. 9.

Csikszentmihalyi, Mihaly. 1990. *Flow: The Psychology of Optimal Experience*. New York: HarperCollins, p. 4.

James, William. 1961. *The Varieties of Religious Experience*. New York: Collier Books, p. 42.

Kluchin, Abby. 2017. At the Limits of Feeling: Religion, Psychoanalysis, and the Affective Subject. In *Feeling Religion*. Edited by John Corrigan. Durham: Duke University Press, p. 257.

Morgan, David. 2017. Emotion and Imagination in the Ritual Entanglement of Religion, Sport, and Nationalism. In *Feeling Religion*. Edited by John Corrigan. Durham: Duke University Press, p. 225.

Proudfoot, Wayne. 1985. *Religious Experience*. Berkeley: University of California Press, pp. 75–78.

Schaefer, Donovan O. 2015. *Religious Affect: Animality, Evolution, and Power*. Durham: Duke University Press, p. 4.

Schaefer, Donovan O. 2017. Beautiful Facts: Science, Secularism, and Affect. In *Feeling Religion*. Edited by John Corrigan. Durham: Duke University Press, p. 70.

Slusher, Howard. 1993. Sport and the Religious. In *Religion and Sport: The Meeting of Sacred and Profane*. Edited by Charles S. Prebish. Westport: Greenwood Press, p. 191.

Taves, Ann. 2009. *Religious Experience Reconsidered: A Building-Block Approach to the Study of Religion and Other Special Things*. Princeton: Princeton University Press, p. 21.

religions

MDPI

Article

Deconversion, Sport, and Rehabilitative Hope

Terry Shoemaker

School of Historical, Philosophical and Religious Studies, Arizona State University, 975 S Myrtle Ave #4302, Tempe, AZ 85281, USA; terry.shoemaker@asu.edu

Received: 8 May 2019; Accepted: 22 May 2019; Published: 27 May 2019

check for
updates

Abstract: This article, based on qualitative interviews and ethnographic research, explores three types of on-the-ground rehabilitative hope supplied by sport for many post-evangelicals within the upper Bible Belt region traversing through the process of deconversion. First, sport is an often-cited space that is identified as broadening social networks, leading to initial questioning of inherited religiosity. Second, sport offers a level of amelioration of relational fissures caused by religious shifts away from evangelicalism. Last, this research indicates that post-evangelicals highly value spaces for discussions of social justice, and athletic activism offers symbolic solidarity. Thus, sport and deconversion can be intertwined for Southern post-evangelicals. In the end, I argue that the triangulation of deconversion, hope, and sport within a Southern context creates a way of understanding the changing Southern ethos and pathos demarcated by a shifting away from a conservative Protestantism historically dominant in the region.

Keywords: deconversion; sport; hope; bible belt; religion; rehabilitation

---

1. Introduction

In 2016, I began conducting ethnographic research to analyze changing forms of religiosity within the region of the United States known as the Upper South or the northern portion of what is generally referred to as the Bible Belt. The target respondents for this research were those once acculturated into conservative Protestantism, and my interview guide was built to discover patterns and themes in the lived experiences of how one socially and subjectively navigates deconversion from a pervasive host religious tradition in the South. The interview guide consisted of questions about religious upbringing, factors in deciding to shift religiosity, and how people continue to socially negotiate religiosity in their present life. Although the semi-structured interview guide consisted of no questions regarding sport, I was surprised to discover that sport was introduced to the discussion by approximately 15% of the consultants of the project.[1]

Sport plays a special role in the Southern ethos and praxis. In a previous collaboration, Eric Bain-Selbo and I analyzed and theorized this affinity toward sports in the South (Bain-Selbo and Shoemaker 2016). Our argument focused on college football and stock car racing as a functional means for Southerners to perpetuate their regional identities and construct sacrality. In this way, we proposed sport and religion competed as the primary repository for preserving the Southern heritage. In this paper, I take a slightly different approach to understanding the intersection of sport and religious deconversion in the Bible Belt through offering a finer grained portrait of the lived realities of some Southerners deconverting from conservative Protestantism (e.g., Evangelicalism, Fundamentalism, Pentecostalism). Through illuminating three ways that sport integrated into the

---

[1] Although 15% might constitute a small percentage of the overall respondents, this number represents the number of respondents who referenced sport without any provocation. The importance of sport, as illustrated throughout this paper in the deconversion process, could be much higher had the project intentionally been investigating sport and deconversion.

discussions of deconversion and newly forming spiritualities during that ethnographic data gathering, I will argue that sport can offer a rehabilitative hope for familial fracturing caused by religious deconversion, a spatial opportunity in a restrictive milieu to expand one's social network, and a space to align with contemporary social issues. In each of these areas, some consultants pointed to sport as instrumentally rehabilitative in their development and perseverance. As a means of illustrating these instances, I include multiple vignettes throughout the paper. The broader relevance of this paper is to demonstrate that sport plays particularly significant roles in the lives of individuals that intersect with deconversion and to provide some research directions for future projects regarding sport and religion.

2. Religion and the Bible Belt

Data on religious affiliations in the United States indicate that there is an increasing trend toward Spiritual But Not Religious (SBNR), a decrease in traditional forms of religiosity, and a common practice of switching religious affiliations (Jones and Cox 2017). The dynamic nature of subjective religiosity and the numerous options that Americans have in the religious marketplace are well documented (Smith et al. 2015; Stark and Finke 2000). Yet, as Elizabeth Drescher argues, social environments must be considered when studying religious affiliations. As Drescher states, "changes in religious affiliation, while not determined by cultural factors alone, are enabled or constrained by culture" (Drescher 2016, p. 56). In some regions of the United States, like the Bible Belt, the social constraints tend to exert social, political, and religious pressures on how individuals religiously affiliate and practice. Each of these factors—fluidity, social constraints, and autonomy—form a nexus for understanding religion in America today. The numerous religious market options available and social pressures can complicate these shifts, and this can prolong the process of religious shifting. Much of what scholarship has concentrated on is the quantifying of the shift from A to B, but there can exist several subpoints along the subjective trajectory from point A to B. The moments, decisions, actions, and choices between the originating position and the landing position are the focus of this study.

Because this study seeks to analyze "phenomena not as detached things-in-themselves, but as connected things-in-the-world" (Rogers 2012, p. 10) and "the making of identities in highly heterogeneous and fast-changing social contexts" (Altglas 2014, p. 475), the initial research project assembles and analyzes multiple forms of data and sources including qualitative interviews, participant observations, and regional histories. This form of research, referred to as a *bricolage*, is a type of ethnographic research useful in projects in which the researcher must assemble from various resources the emergence of cultural transitions. This specific portion of the research draws heavily upon the 65 qualitative interviews conducted over two years. For this paper, these interviews were recorded, transcribed, and analyzed seeking recurring themes and patterns related to sport and religious deconversion, while also considering the conditions of the cultural milieu.

The following vignette of a consultant that I call Mark illustrates how Mark's religious shift is complicated and constrained by his social context, which has played out over time.[2]

Vignette 1

Mark is white, in his thirties, and identifies today as SBNR, although he still attends a progressive Christian church with his wife. Mark was proud to describe growing up in North Carolina. His family was close, and he regularly participated in his local church, attended by several of his family members including his parents and grandparents. Their conservative Protestant church cultivated shared values, rituals, and heritage for their family and supplied them a weekly gathering time to commune followed by a family meal. In his adolescence, Mark explained that he made the decision to live his life in "close proximity to the church." This decision was soteriologically and socially based.

---

[2] To protect the identity of my consultants, I utilize aliases. In addition, the locations of some narratives have also been modified as an additional layer of anonymity.

There came a time during college when Mark began to question his religious beliefs and values. The foundational conservative Protestant beliefs of atonement theology, scriptural inerrancy, and divine plausibility began to rapidly crumble as he studied philosophy and existentialism. Today, more than a decade into his faith deconversion, Mark still has not been able to share authentically his views on religion with his family or that he self-identifies SBNR. He explains that divulging this information to his family would be too devastating for his grandparents particularly. Instead, he, his wife, and friends meet with a small, more progressive Christian community that offers more freedom in discussing religiosity and spirituality. This permits Mark and his wife an opportunity to appease his family and remain within social expectations, but also remains a point of contention within his family because he does not attend the family church. Taking the social import of religious affiliation and participation in the South into consideration explains partially why Mark hesitates to reveal his perspectives and opinions to his family. Revealing his religious skepticism would be perceived as an equivalency of renouncing his family heritage.

In the Bible Belt region, conservative forms of Protestantism are the dominant religious affiliations (Smith et al. 2015). The varieties of Protestant denominations physically dominate the landscape with thousands of churches, private Christian schools, parachurch organizations, and signage. This is a point of pride for many Southerners, and I have heard it debated which area or Southern state composes the "buckle of the Bible Belt" or that which idealizes the Protestant dominance. Beyond affiliations and infrastructure, the religious history of the Second Great Awakening creates an imagined exceptionalism in that the divine would choose this region to begin religious revivals. In addition, Protestant churches remain heavily segregated today (Emerson and Smith 2000), and the legacy of the Civil War forms another informative layer with the "religion of the Lost Cause", slavery, and Jim Crow (Wilson 1980). Often, the proud heritage of many Southerners is located with a specific church where family lineage can be traced. Many Southerners retain the practice of attending church together on Sundays followed by a large, post-gathering meal typically held in the patriarch's home, although like the rest of the United States, religious attendance is declining (Jones 2014).

Unlike Mark's example, many of my consultants do decide to reveal their religious doubts and critiques of their home church and, more broadly, of Southern Protestantism. This leads to tremendous family and social fracturing that can be quite devastating. Many of the consultants of my project join in a movement that scholarship is now calling "Emerging Christianities" (Moody and Reed 2017; Bielo 2017). Marti and Ganiel (2014) suggest that these communities share the continual practice of deconstruction, a process of critically analyzing one's inherited cultural and religious values, forms, and practices. Philip Harrold recognizes this in his study of Emerging Christianities' online resources (Harrold 2006). He discovers that "individuation—the conscious effort to choose one's personal identity and core convictions," is at the crux of these communities (Harrold 2006, p. 79). There exist then site-specific distinctions between these communities due to the inherited acculturations that are being deconstructed. Because religiosity and spirituality are considered in flux, these communities permit a freedom in cultivating subjective spiritual or religious orientations.

These Emerging Christian communities invert Davie (1994) notion of "believing without belonging"—the idea that many people maintain religious beliefs but choose to not be part of a traditional religious community with established gathering spaces and times. Instead of "believing without belonging," my research finds that the Emerging Christian communities are spaces of "belonging without believing." Doctrines and creeds are decentered and less important. Stated otherwise, confessional assent or doctrinal alignments are not required for membership. What I discovered was that these communities in the Upper South serve as spaces for former conservative Protestants to renegotiate their religiosity and/or spirituality (the term often utilized by the research consultants).

In essence, my research argues that in the Upper South, the Emerging Christian communities provide a space for ex-conservative Protestants to reconstitute their Self, a reconstitution process that is understood as a religious practice. This religious practice extends into multiple aspects of individual and collective life. My project consultants utilize their communities to grapple with various forms of identities like national, regional, sexual, gender, and race and the intersectionality of these identities. The community is a central space for reforming subjectivity during deconversion, and these communities offer an in-the-moment space that rehabituates one into new norms.

Barton (2011, 2012) analyses of the Bible Belt argue that there exists a compulsory Christianity that is part of a more oppressive Bible Belt panopticon attempting to regulate those within its surveillance boundaries. Barton's work focuses specifically on how sexuality is surveilled within the Upper South region through the dominance of the Christian panoptic mechanisms. My research corresponds with Barton's and discovers that multiple aspects of individual identity are regulated within the region. For instance, along with heteronormativity, patriarchal and racial norms are imposed through the panoptive infrastructures. Barton's work helps in understanding why such tension is created with the continued practices of the Emerging Christianities in permitting individual subjectivities to form. This pushes not only against the religious norms but also against strongly maintained cultural practices.

My project's consultants described the rigidity and oppressiveness of the region in various terms, but almost all referenced the difficulties they have had in shifting away from the religion of their youth. For instance, one consultant describes her life in the South as a place where "you can't be yourself. You have to watch what you say. You can't be an individual. There's always someone watching and judging." Several others echoed these sentiments, and these identity shifts often result in the fracturing of familial and social ties. Furthermore, the new spirituality produces new values and political perspectives. This creates individual perspectives and practices that are often at odds with the dominant religio-political climate of the region. Also, importantly, in most of these instances, the project consultants remain in the Upper South and must still learn to navigate the Bible Belt context as religious deconverts.

Although most consultants pointed to their religious community as the primary space for the work of deconversion, some of the project consultants indicated sport serves as an important aspect of their deconversion practices and strategies. The references to sport emerged in three categories which are explored below: Early social network expansion, a safe space for continued family relationships, and alignment with athletes' social justice proclamations. In each of these areas, sport is attributed a quality of rehabilitative hope.

3. Social Network Expansion through Sport as Rehabilitative Hope

Sport provides many benefits for the formation of youth across America. This obviously includes physical development but also encompasses the development of self-identity and relational network formation (Bruner et al. 2017; Côté and Fraser-Thomas 2016). Many youth maintain positive perspectives of sport participation and school athletic programs (Lubans et al. 2011). This is certainly not always the case, as there are reported bullying and overtraining issues that can emerge in some instances (Meyer et al. 2015). Less analyzed are the ways in which team and individual sports provide an opportunity for participants to expand their social networks and the effects of these social network expansions.

Playing sports includes social interaction. This involves multiple layers of interactions occurring simultaneously: coaches and athletes, teammate with other team members, athletes with referees, family members with athletes, athletes and opposing team members, etc. Negative interactions like fan feuds and violence are explored typically by scholarship (Lewis 2007; Wann et al. 2017). However, the time commitments involved formally and informally with practices, games, travel, and off-field activities like player recognitions lead to opportunities for expanding social networks and developing long- and short-term relationships. Recent scholarship offers a way of thinking about informal interactions between players, but few studies aim to analyze the long-term effects of youth sport participation vis-à-vis social interactions (Erickson and Côté 2016).

As it pertains to social interactions in college sports in the American South, Eric Bain-Selbo argues that aspects of Southern fandom—i.e., tailgating, shared social identities, spectatorship, etc.—equate to

Victor Turner's *communitas* (Bain-Selbo 2012, pp. 27–52). His work locates the social significance of college sports fandom as a liminal state of being that separates fans from the mundane (Bain-Selbo and Sapp 2016). Some of Bain-Selbo's fan interlocutors indicate that they feel more connected to the college football fan community than their religious community. In addition to the emotive aspects of the fan experience, many Southerners also understand sport as providing a place for masculinity to be cultivated (Dowland 2011; Miller 2002). In a modern world that many Southerners perceive as emasculating, sport offers an opportunity to demonstrate strength and ability, while reifying the traditional patriarchal roles and privilege.

Furthermore, many conservative religious communities in the South create insular social networks. These networks elevate the importance of the religious community over external forms of social engagement. Robert Putnam argues that religious communities promote healthy levels of broad volunteerism in America, except in the case of evangelicals who focus more on internal forms of volunteerism (Putnam 2000, pp. 75–77). In such cases, religion and sport are often seen as rivals competing for time and monetary commitments from congregants and/or fans (McMullin 2013). To remedy this competition, some religious communities offer their own sporting leagues. This supplies the participatory sporting demands from some congregants and reduces time spent building social networks beyond the boundaries of the church institution. Moreover, because of the maintained objective of evangelizing, many Christians perceive sports as a place to proclaim and share one's religious faith (Hoffman 2010).

To summarize, there often exists a tension between sport and religious commitments in the South. On the one hand, sport is understood as an opportunity for discipline, evangelism, and further social interactions of the religious community. On the other hand, sport can interfere with religious participation and practice.

During qualitative interviews, a vital question for the research was to locate in the consultants' viewpoint when and where the consultants first began to question their inherited religiosity. My assumption was that forms of intellectual doubt emerged as the consultants increased in educational attainment or that a particular traumatic experience provoked the religious shift. In most cases, these were the significant factors that were highlighted by the consultants. However, in some cases, sports participations, specifically basketball, in the South resulted in project consultants questioning their faith. These cases demonstrate how social interactions in sporting spaces can disrupt individual lives religiously and prompt a spiritual deconversion in a milieu that attempts to regulate religiosity. Two particular cases demonstrate how sporting interactions can be religiously disruptive.

Vignette 2

Todd "grew up hard, red clay Baptist Fundamentalist ... in an all-white, clustered fundamentalism" in Tennessee. Although as a young boy he enjoyed the certainty and homogeneity of his community, in high school, he began to question the racial norms of his church community. These questions emerged when he was exposed to different social and racial classes while playing with his local high school basketball team. After busing policies were changed in his hometown, Todd's public school's demographics significantly changed, increasing the number of people of color. During his junior year, Todd was the only white player in his basketball team. He recalls, "Because I was an athlete, I was exposed to a different social class through athletics. From traveling to play basketball to football, I encountered people that were outside the rural bubble." Reflecting back to his basketball experience as a player, he notes that it was the social interactions with his teammates that radically reoriented his outlook on individual and systemic racism, which eventuated also in his initial questioning of his religious community that was demographically an all-white community. Todd explains, "Those things began fraying at my worldviews that I'd been given. I was aware that there were big issues and that the church could be desperately wrong on those issues."

Vignette 3

Like Todd, Lisa's upbringing was heavily tied to her church community, which was a conservative, southern Church of Christ. She and her siblings were homeschooled as a means of protecting them from secularizing influences. She recounts that her church and family constituted the overwhelming majority of her social network as a child. Lisa thinks that her current spirituality started as she questioned her faith "in layers." The initial space for this questioning was at a homeschooling basketball league with other conservative Christian youth. Her parents helped form the homeschooling league as an athletic alternative to local non-religious leagues. During this time, though, she encountered Christians who maintained differing views from her family. It was in this instance that she recognized the diversity within the Christian tradition, which created a relativizing momentum that has progressed to today. This league continued to alter her perspectives when, later in life, one of those homeschooling teammates came out as gay. Today, Lisa proudly remembers these encounters with other youth participants as significantly informative to her current spirituality.

These two examples indicate that, in some cases, social interactions with teammates can have a profound impact on subjectivity. These instances align with Peter Berger's proposal that the modern world's religiously pluralistic society relativizes subjective perspectives on religion and spirituality (Berger 2014). This relativizing, argues Berger, "becomes a permanent experience" in the modern world (3). Thus, the conditions of modernity produce lived realities that are in constant navigation of a multitude of ideas. Berger's work focuses primarily on the ways in which religious pluralism interacts with subjective religiosity, but the continual negotiation of religious pluralism extends outside of religious institutions into sporting spaces. These spaces, and the social interactions that occur, can have similar relativizing effects. Even in the South, where some conservative, religious families attempt to build insular communities, sporting interactions can disrupt the imagined insularity.

As a number of the project consultants reflect back on their deconversion, they locate the originating point connected to sport participation. Each of these consultants attributes a rehabilitative hope to sport in this way. As often is the case, the present deconversion is understood as a distinctively improved position relative to the previous religious commitments. While a portion of consultants continue to reform their Self, sport is retroactively perceived as the factor which propelled them on their current journey. Thus, sporting participation is credited with the initial decoversion move. The deconversion is understood as humanistic in the sense that human connection significantly informs decisions and life trajectories. Thus, in some cases, sport and deconversion are intertwined in forming the reconstituted Self.

4. Family Tensions, Sport, and Rehabilitative Hope

The importance of familial religiosity can result in both negative and positive consequences for individuals. Studies on these consequences analyze the correlations between familial religiosity and disciplinary practices (Bartkowski and Wilcox 2000), educational attainments (Danso et al. 1997), and sermon influence on familial life (Gershoff et al. 1999). In some studies, shared, familial religious practices are posited as a cohesive mechanism (Marks et al. 2017; Chelladurai et al. 2018; Dollahite and Marks 2009). In applied scholarship, social workers are recommended to be more attentive to familial religion and spiritual practices (Godina 2014).

In this study, as the journey formulates after religious deconversion, tensions tended to emerge with social networks that were essential in consultants' past. This is a common theme that manifested throughout my research. Although social network fissures can occur in dating or business practices, the most common reference to social fracturing was with parents and siblings. The following are just a small sampling of how consultants describe their relationship with their families after deconversion:

"It's a weird thing. I love my family. But how it used to be and how it is now, it's totally different … The reason I don't like to talk about religion is because I've seen what happens when you talk about it. When you get to a point with evangelicals or people who hold strongly to a worldview that is so ingrained in my family they are only here, to them, this is like purgatory. When you bring up things that you don't believe in, my parents are like that's our entire identity."—Chris, twenties, Kentucky

"I'm in a closet [about spiritual identity] and I don't know how to come out to my parents . . . I don't go to my parent's house right now."—Denice, fifties, North Carolina

"I hold [my parents] at arm's length. I don't tell them the details of my life."—Brandy, thirties, Kentucky

"There's just a lot that you'll never see eye to eye on. It's not worth me trying to change [my parents'] minds on anything. I don't want to do that. We just talk about surface-level topics now."—Joan, thirties, Tennessee

In some cases, relationships were completely severed, and ostracizing occurred. This prompted some consultants to relocate geographically, but in most cases the consultants attempt to maintain relationships with family members although this requires ongoing work and negotiation. My data capture the divisions as the consultants describe religious interventions, family visits intentionally planned around religious holidays, and multiple attempts to recruit them back to the host church. These moves typically result in fracturing family relationships more rather than successfully healing the fissures.

Familial tensions and the Southern ethos often produce a tacitly held agreement to privatize religious discussions for the sake of peace. In these situations, shared commitments and allegiances are sought after to replace the loss of religious connections. This is certainly true in the earlier mentioned narrative of Mark. He describes his shared commitments to college basketball as central to maintaining his family ties. The moments of shared fandom become significantly important as a safe space for familial interactions. For instance, Mark's interpersonal conflicts with his parents and grandparents are often placated through shared allegiances to the University of North Carolina basketball and football teams. Their shared fandom supplies a space for the family to temporarily bracket other tensions and enjoy a unifying cause. Mark refers to the moments of spectatorship as sacred.

Vignette 4

Like Mark, Jack's family ties, particularly his relationship with his father, suffered after he revealed that he no longer maintained his beliefs in the doctrines of their Free Will Baptist church. Growing up in the eastern part of Kentucky, the family's life centered within the religious community. Yet, as Jack's deconversion occurred, he found himself often at odds with his father in religious discussions. Jack's father continues to push Jack to return to the faith of his youth, but Jack has now relocated to Tennessee and is part of a progressive Christian community.

As Jack described his relationship with his father, it was their commitment to West Virginia University's college football team that provided a remediating space to positively connect with his dad. "Here, on this couch watching football, that's our best times now. We tend to argue about religion and definitely about politics, but college is where we can set that aside and just be father and son again."

In each of these situations, religious difference is central to the familial tensions. For Mark and Jack's family, sport remediates an intra-family religious conflict that is occurring and ongoing. There appears to be no end in sight for their religious disagreements. Scholarship posits that sport can play a significant role in remediating social conflict at macro levels (Giulianotti 2011; Majaro-Majesty 2011), yet less research analyzes how sport interferes or remediates inter-family conflict (Simmons and Greenwell 2014). Where religion previously functioned as a cohesive social space, sport now functions to fill that void. It is not simply the case that sport fandom is a distraction from an ongoing conflict. In Jack and Mark's experience, sport now accomplishes for their family what their previously shared religiosity did. With less shared experiences, the families tend to sacralize the shared time with sporting spectatorship and fandom. In this way, there exists a rehabilitative hope to the shared sports fandoms.

Sport displacing religion in this way should not come as a surprise. The Southern ethos prides itself on passing on a cultural heritage. The cultural heritage includes several aspects of the Southern way of life, like food culture, geographical embeddedness, sports, and religiosity. Historically, this heritage transmission has been facilitated by the rootedness and intricacies of the localized kinships and the home (Allen 1990). As more Southerners position themselves religiously distinct from their parental inheritance (Thompson 2013), it is reasonable to imagine that other aspects of that shared history would take on a more significant meaning for the families. Thus, if religion becomes less relevant as a conduit of shared meaning-making, families will seek out other shared connections, like sports fandom, to fill the voids left by a lack of common religiosity.

5. Social Justice, Sport, and Rehabilitative Hope

Like the first two, the third type of rehabilitative hope supplied by sport during processes of deconversion in the South is relational in nature. To briefly recap, some consultants of the research project designated team participation in sports as the initial site for religious and spiritual expansions. The social relationships cultivated on the court of basketball facilitated a broader questioning of inherited religious norms that eventuated in deconversion. As a means of maintaining relationships after the initial deconversion process, for other consultants, sport serves as a shared space for bonding and transmitting culture. The third type of hope involves the development of a different kind of morals, values, and relational alignments accompanying decoversion.

A key aspect of deconversion highlighted in the initial research study was the ways in which the consultants drastically shifted moral positions after deconversion. This included religious values, but also political reorientations. Although most of my consultants were raised in conservative, politically leaning households, a majority of consultants describe themselves today as progressive, Democrat, or Independent party-leaning. Perspectives on several issues like civil marriage, abortion, and gun control were all noted as subjects about which values have been reoriented. The modified values significantly add to familial tensions that are already existent due to deconversion. During data collection, sport also emerged as a relational space to discuss, relate, and find solidarity with others who shared similar values.

Vignette 5

Take, for instance, Kasey and Evelyn, a married couple who have journeyed on their deconversion processes together. Kasey admits that he is more orthodox, while Evelyn describes herself as more nontraditional. During their search for a religious community, the two thought that they had found a church "on our terms." Kasey explains, "We wanted the possibility of a woman in leadership, a church that cared about the environment and directed their gaze beyond themselves. We wanted a church to at least acknowledge that there are things happening in the world." Their newly discovered values modified what they sought after in a religious community.

While in Kentucky, they found a church but were quickly disgruntled when they discovered that Evelyn would not be permitted to become a member because she taught a yoga class in the community. As she describes it, her yoga class was separated from any religious connections, but the church explained that teaching yoga was at odds with their religious beliefs and practices. Evelyn thinks that the real issue was that the yoga class provides her with an expanded social circle beyond the church community. She admits, "most of my friends are not Christians. I have more in common with people of varying faiths than most

Christians around here. I just have more in common with them." As the spiritual journey of deconversion developed for Evelyn, she discovered that the exercise of yoga provided the opportunity to connect with other similar people.

The shifts in values and political orientations push many consultants into an active political mode opposite to that of their past. Qualitative interviews for the research project occurred during the presidential campaigning and election of 2016, and several of the project consultants were politically mobilizing and marching. Many of the consultants were raised in racially segregated religious communities in the South. Many of them attended predominantly white, religious school systems. Today, a majority of the consultants are active in allying with people of color, non-heterosexuals, and other marginalized people in the Bible Belt region. I discovered in the qualitative data an attempt to repair relationships with people at the margins of the Bible Belt society. These interactions with marginalized people groups tended to disrupt previously held notions of patriotism. I described this as a shift from a civil religious nationalism to a critical patriotism. When I asked the consultants about their current values and political positions, I typically received responses that argued for transcending national identity for a more globalized understanding of human connections.

Vignette 6

Chad was born outside of the Bible Belt, but his family relocated to Tennessee when he was a small child. Quickly, Chad explains, his family joined a local Protestant group affiliated with a Pentecostal denomination. He described several religious experiences as a teenager with his religious community, but his religiosity started when he began schooling at a Bible College. It was at the college that Chad began to question the parameters and foundations of his childhood faith. Today, he is vocal about his political leanings and is active with a political mobilizing organization. As Chad posits, "The ways that I'm patriotic are in the ways that I still have the right to say, 'I am not fucking patriotic. This is absolute garbage.' That sense of freedom and those rights, that's how I'm patriotic. Otherwise, I'm about human rights, which transcends our borders." As Chad continues to explain his current political perspectives and allegiances, he aligned his positions with Colin Kaepernick's, the quarterback athlete activist who refused to stand for the national anthem at football games. As Chad explains, "I kneel in solidarity with Kaepernick. This notion drives my family absolutely crazy."

Those on the deconversion journey appear to seek out prophetic voices that articulate their sensibilities. Many consultants referenced Martin Luther King, Jr. and the Civil Rights Movement as an ideal way that religious perspectives can motivate political mobilization. Elsewhere, I have argued that political activism by athletes today constitutes a form of prophetic activity (Shoemaker 2019). Like Chad, other consultants referenced NBA athlete activism as indicative of their current political stances and directions. To be more specific, project consultants referenced athletic activism working towards improving race relations in the United States as the model for their political aspirations. When asked about current issues within the Bible Belt, a majority of consultants indicated that racial divisions and systemic racism continue to plague the region. In this way, athletic activism offers a resonating message for the project consultants. Thus, the spiritual formation during the deconversion process seeks prophetic voices outside of traditional religious institutions.

This third type of hope grounded in athlete activism potentially challenges the second. If sport is a shared space for family, politically charging the sporting spaces might reactivate the religio-political tensions that families are seeking to avoid.

6. Conclusions: Sporting Hope and the Changing South

During periods of deconversion from demanding types of religious commitments that are heavily tied to social networks, those deconverting might seek to find alternative spaces to cultivate their new forms of spirituality and religiosity. As might be expected, new forms of spiritual and religious communities emerge to fill those voids. What is less expected, and thus less examined, are the ways that other cultural phenomena and spaces outside of traditional forms of religion also aid in this religious and spiritual exploration. This corresponds with Hutch (2012) premise that we should seek to "articulate the lived experiences of the individual, notwithstanding the presence or absence of a commitment to a particular set of religious beliefs and practices" (p. 141). In other words, religious beliefs and practices, and in this case, religious deconversion, exist alongside various other lived experiences. People are continuously assessing and integrating these experiences to formulate subjectivities, positions, and perspectives. This is a generally shared reality beyond the Southern milieu. More work needs to be conducted to assess the particulars of deconversion in other areas of the United States and beyond that can incorporate the privatization of religion for the health of relational networks, the navigation of increasingly pluralistic ideas and perspectives, and modes of patriotism.

The sample size of this study is too small to make any substantial inference. What the data do suggest is that sport *can* play an integral role in the deconversion processes and, for some individuals in the South, sport aids in filling particular relational voids. Individuals on deconversion journeys in the modern Bible Belt are not constrained by the boundaries of religious institutions or traditions. As lives are reconstructed, individuals discover resources for reconstituting their Self. These resources are typically understood as having a sacred quality for those individuals. In this case, this reconstitution work includes mining the past to question where it all began, reflecting on how the present conditions are ameliorated or on future oriented methods of challenging the structural systems. In each of these realms, sport can be a contributing space for discovering rehabilitative hope in these situations.

These individual deconversion changes reflect broader social transformation in the Bible Belt culture. In the American South, change has been viewed with intense suspicions. Yet, like the rest of the country, the Bible Belt is experiencing transformations in various sectors of life like economics, demographics, and religion. Many Southerners attempt to resist these changes and seek to rediscover an ever-disappearing past. However, these "efforts to insist on a return to traditional pieties thus inevitably clash with the structure of the modern economy and produce recurring cries of moral crisis" (Cahn and Carbone 2010). The moral crises are typically articulated in religious and political spaces in the South. Sport, especially college sports, currently remains unscathed from the moral crisis discourse. In this way, sport offers a safer space for families and friends to maintain their relationships and broaden their social networks.

Moreover, the South is still plagued by racialized structures. Religious communities, broadly speaking, have failed to racially integrate society and help move the Bible Belt forward in a healthier, racially integrated direction. Churches are typically still racially segregated, but college sports, which are extremely important for many Southerners, offer an opportunity for more racial and gender integration due to Title IX mandates. In this way, secularized, college sporting institutions serve as a model for those deconverting to more progressive forms of religiosity and spirituality.

Southerners also seek mechanisms by which to transmit their heritage to their children and grandchildren. Whereas the local church has been central in these endeavors in the past, today many Southern families simply do not share religious values, ideas, or institutions. This forms a need to seek other avenues for cultural transmission. Sport is one site displacing religion in this way. Sport offers a way of transmitting shared values like team fandom, identities, and practices. This assuages the loss of common religiosity and family fracturing due to this loss. In other words, where religion serves as a site for familial conflict, sport tempers this reality and offers a way to connect and pass on heritage. In a previous work, I described sport as an institutional "mason jar" that preserves the ethos and pathos of Southern culture (removed for peer review). In this way, sporting commitments might become increasingly important in the future of the South.

Yet, the mason jar strength of Southern sport could be jeopardized by athletic activism. This is neither a defense nor a critique of athlete activism, but more a point of how activism by athletes, whether college or professional, could disrupt how sport is utilized as a rehabilitative hope. If sport becomes politicized, then the personal aspects of remediation might fade. Many Southerners would understand athletic activism as a corruption of the purity of the athletic fields and courts. This only serves to support the notion that in the Southern mind, sport possesses a sacred quality that is separate from the mundane sector of politics.

As the United States continues to modernize and secularize, regional identities are becoming less distinct. With increase access to mobilization, fewer people remain anchored geographically. These relocations pluralize regions, creating a new heterogeneity. In order for the regional identity of "Southerners" to proceed into the future, enough shared common interests in perpetuating the regionalism must exist. It is my argument herein that sport is the site for this regional cultivation and perpetuation. In fact, sport is a site for rehabilitative hope for Southerners today.

Funding: The research received no external funding.

Conflicts of Interest: The author declares no conflict of interest.

References

Allen, Barbara. 1990. The Genealogical Landscape and the Southern Sense of Place. In *Sense of Place: American Regional Cultures*. Edited by Barbara Allen and Thomas J. Schlereth. Lexington: The University of Kentucky Press.

Altglas, Veronique. 2014. 'Bricolage': Reclaiming a Conceptual Tool. *Culture and Religion* 15: 475–93. [CrossRef]

Bain-Selbo, Eric. 2012. *Game Day and God: Football, Faith, and Politics in the American South*. Macon: Mercer University Press.

Bain-Selbo, Eric, and D. Gregory Sapp. 2016. *Understanding Sport as a Religious Phenomenon*. London and New York: Bloomsbury Academic.

Bain-Selbo, Eric, and Terry Shoemaker. 2016. Southern reconstructing: Sport and the future of religion in the American South. In *Sport and Religion in the Twenty-First Century*. Edited by Brad Schultz and Mary Lou Sheffer. Lanham: Lexington Books.

Bartkowski, John P., and W. Bradford Wilcox. 2000. Conservative Protestant child discipline: The case of parental yelling. *Social Forces* 79: 265–90. [CrossRef]

Barton, Bernadette. 2011. 1CROSS + 3NAILS = 4GVN: Compulsory Christianity and Homosexuality in the Bible Belt Panopticon. *Feminist Formations* 23: 70–93. [CrossRef]

Barton, Bernadette. 2012. *Pray the Gay Away: The Extraordinary Lives of Homosexuals in the Bible Belt*. New York: New York University Press.

Berger, Peter. 2014. *The Many Altars of Modernity: Toward a Paradigm for Religion in a Pluralist Age*. Boston and Berlin: Walter de Gruyter.

Bielo, James. 2017. The Question of Cultural Change in the Scientific Study of Religion: Notes from the Emerging Church. *Journal for the Scientific Study of Religion* 56: 19–25. [CrossRef]

Bruner, Mark W., Shea M. Balish, Christopher Forrest, Sarah Brown, Kristine Webber, Emily Gray, Matthew McGuckin, Melanie R. Keats, Laurene Rehman, and Christopher A. Shields. 2017. Ties That Bond: Youth Sport as a Vehicle for Social Identity and Positive Youth Development. *Research Quarterly for Exercise and Sport* 88: 209–14. [CrossRef] [PubMed]

Cahn, Naomi, and June Carbone. 2010. *Red Families v. Blue Families: Legal Polarization and the Creation of Culture*. Oxford: Oxford University Press.

Chelladurai, Joe M., David C. Dollahite, and Loren D. Marks. 2018. The family that prays together . . . : Relational processes associated with regular family prayer. *Journal of Family Psychology* 32: 849–59. [CrossRef]

Côté, Jean, and Jessica Fraser-Thomas. 2016. Youth involvement and positive development in sport. In *Sport and Exercise Psychology*. Edited by Peter R. E. Crocker. Toronto: Pearson, pp. 256–87.

Danso, Henry, Bruce Hunsberger, and Michael Pratt. 1997. The role of parental religious fundamentalism and right-wing authoritarianism in child-rearing goals and practices. *Journal for the Scientific Study of Religion* 36: 496–511. [CrossRef]

Davie, Grace. 1994. *Religion in Britain since 1945: Believing without Belonging*. Oxford and Cambridge: Blackwell.

Dollahite, David C., and Loren D. Marks. 2009. A conceptual model of family and religious processes in highly religious families. *Review of Religious Research* 50: 373–91.

Dowland, Seth. 2011. War, Sports, and the Construction of Masculinity in American Christianity. *Social Compass* 5: 355–64. [CrossRef]

Drescher, Elizabeth. 2016. *Choosing Our Religion: The Spiritual Lives of America's Nones*. New York: Oxford University Press.

Emerson, Michael O., and Christian Smith. 2000. *Divided by Faith: Evangelical Religion and the Problem of Race in America*. Oxford: Oxford University Press.

Erickson, Karl, and Jean Côté. 2016. An Exploratory Examination of Interpersonal Interactions between Peers in Informal Sport Play Contexts. *PLoS ONE* 11: e0154275. [CrossRef]

Gershoff, Elizabeth Thompson, Pamela C. Miller, and George W Holden. 1999. Parenting influences from the pulpit: Religious affiliation as a determinant of parental corporal punishment. *Journal of Family Psychology* 13: 307–20. [CrossRef]

Giulianotti, Richard. 2011. Sport, peacemaking and conflict resolution: A contextual analysis and modelling of the sport, development and peace sector. *Ethnic and Racial Studies* 34: 207–28. [CrossRef]

Godina, Lidija. 2014. Religion and parenting: relationship ignored? *Child & Family Social Work* 19: 381–90.

Harrold, Philip. 2006. Deconversion in the Emerging Church. *International Journal for the Study of the Christian Church* 6: 79–90. [CrossRef]

Hoffman, Shirl James. 2010. *Good Game: Christianity and the Culture of Sports*. Waco: Baylor University Press.

Hutch, Richard. 2012. Sport and Spirituality: Mastery and Failure in Sporting Lives. *International Journal of Practical Theology* 5: 131–52. [CrossRef]

Jones, Robert P. 2014. Southern Evangelicals Dwindling—and Taking the GOP Edge with Them. *The Atlantic*, October. Available online: http://www.theatlantic.com/politics/archive/2014/10/the-shrinking-evangelical-voter-pool/381560/ (accessed on 8 March 2019).

Jones, Robert P., and Daniel Cox. 2017. America's Changing Religious Identity: Findings from the 2016 American Values Atlas. Public Research on Religion Institute. Washington, D.C. Available online: https://www.prri.org/wp-content/uploads/2017/09/PRRI-Religion-Report.pdf (accessed on 8 March 2019).

Lewis, Jerry M. 2007. *Sports Fan Violence in North America*. Lanham: Rowman & Littlefield Publishers.

Lubans, David R., Philip J. Morgan, and Ann McCormack. 2011. Adolescents and school sport: The relationship between beliefs, social support and physical self-perception. *Physical Education and Sport Pedagogy* 16: 237–50. [CrossRef]

Majaro-Majesty, Henry. 2011. Ethnicity, conflict and peace building: Effects of European football support in Nigeria. *Soccer & Society* 12: 201–11.

Marks, Loren D., Trevan G. Hatch, and David C. Dollahite. 2017. Sacred practices and family processes in a Jewish context: Shabbat as the weekly family ritual par excellence. *Family Process* 57: 448–61. [CrossRef]

Marti, Gerardo, and Gladys Ganiel. 2014. *The Deconstructed Church: Understanding Emerging Christianity*. Oxford: Oxford University Press.

McMullin, Steve. 2013. The Secularization of Sunday: Real or Perceived Competition for Churches. *Review of Religious Research* 55: 43–59. [CrossRef]

Meyer, Gregory D., Neeru Jayanthi, John P. Difiori, Avery D. Faigenbaum, Adam W. Kiefer, David S. Logerstedt, and Lyle J. Micheli. 2015. Sport specialization, Part I: Does early sports specialization increase negative outcomes and reduce the opportunity for success in young athletes? *Sports Health* 7: 437–42. [CrossRef]

Miller, Patrick B. 2002. The Manly, the Moral, and the Proficient: College Sport in the New South. In *The Sporting World of the New South*. Edited by Patrick B. Miller. Urbana and Chicago: University of Illinois Press.

Moody, Sarah, and Randall Reed. 2017. Emerging Christianity and Religious Identity. *Journal for the Scientific Study of Religion* 56: 33–40. [CrossRef]

Putnam, Robert. 2000. *Bowling Alone: The Collapse and Revival of American Community*. New York: Simon & Schuster.

Rogers, Matt. 2012. Contextualizing Theories and Practices of Bricolage Research. *The Qualitative Report* 17: 1–17.

Shoemaker, Terry. 2019. *The Prophetic Dimension of Sport*. Cham: Springer International Publishing.

Simmons, Jason, and T. Christopher Greenwell. 2014. Differences in Fan-Family Conflict Based on an Individual's Level of Identification with a Team. *Journal of Sport Behavior* 37: 94–114.

Smith, Gregory, Alan Cooperman, Jessica Martinez, Elizabeth Sciupac, and Conrad Hackett. 2015. America's Changing Religious Landscape. Pew Research Center. Available online: https://www.pewforum.org/2015/05/12/americas-changing-religious-landscape/ (accessed on 8 March 2019).

Stark, Rodney, and Roger Finke. 2000. *Acts of Faith*. Berkeley and Los Angeles: University of California Press.

Thompson, Tracy. 2013. *The New Mind of the South*. New York: Simon & Schuster.

Wann, Daniel L., Paula J. Waddill, Danielle Bono, Holly Scheuchner, and Kristen Ruga. 2017. Sport Spectator Verbal Aggression: The Impact of Team Identification and Fan Dysfunction on Fans' Abuse of Opponents and Officials. *Journal of Sport Behavior* 40: 423–43.

Wilson, Charles Reagan. 1980. *Baptized in Blood: The Religion of the Lost Cause 1865–1920*. Athens: University of Georgia Press.

religions

MDPI

Article

Babe Ruth: Religious Icon

Rebecca Alpert

Department of Religion, Temple University, Philadelphia, PA 19122, USA; ralpert@temple.edu

Received: 24 April 2019; Accepted: 17 May 2019; Published: 23 May 2019

check for
updates

Abstract: Babe Ruth is a mythic figure in American baseball history. His extraordinary skills and legendary exploits are central to the idea of baseball as America's national pastime and are woven into the fabric of American history and iconography. Much has been written about Ruth's life, his extraordinary physical powers, and the legends that grew up around him that made him a mythic figure. The story of Babe Ruth as it has been told, however, has not included its meaning from the perspective of the study of religion and sport. This paper explores the life and legends of Babe Ruth to illustrate the significance of Ruth's identity as a Catholic in early twentieth-century America and the fundamental connections between Ruth's story and the Christian myth and ritual that is foundational to American civil religion.

Keywords: Baseball; Babe Ruth; American Catholicism

---

1. Introduction

Baseball, America's "national pastime," was central to early twentieth-century American mythology. Even if it has been surpassed by football and basketball in the American consciousness and imagination in the contemporary era, baseball serves as a reminder of the virtues and values of the American past. Babe Ruth (1895–1948) was the dominant figure in that narrative. His unparalleled ability to hit home runs farther and with more frequency than anyone before him transformed the game from an institution marred by accusations of gambling and characterized by tough, aggressive play into a demonstration of awe-inspiring power and beauty. Ruth's athleticism combined with his larger-than-life personality changed baseball from an expression of the puritan work ethic located in a pastoral setting to a celebration of the new urban media and consumer culture that would resonate in American life in the 1920s and beyond.

Ruth's status as a mythic hero has been well documented. He has been the subject of numerous works of fiction, documentaries, exhibitions, conferences, biographies, and scholarly articles that demonstrate his importance to baseball history and to American society more broadly (Montville 2006; Leavy 2018; Wehrle 2018; Smelser 1993; Creamer 2005; Keane 2008). Recent scholarship has debunked many of the myths that circulated about Ruth during his lifetime. He was not an orphan. He did not contract syphilis from his profligate ways. In fact, his sexual (and gustatory) exploits were highly exaggerated (Leavy 2018; Wehrle 2018). He neither invented nor profited from the Baby Ruth candy bar that was named, in all likelihood, after him. While scholars have been interested in separating fact from fiction, they have been less concerned with exploring the myths surrounding Ruth to understand the role religion played in his story. While they occasionally resort to the use of religious terminology (saints and saviors, worship and ritual), they have not engaged the concepts of religious studies in understanding his significance. The literature on Ruth has underplayed his role in what some describe as "the religion of baseball." The meaning of Ruth's own religious and ethnic identity as a Catholic in this era also remains unexplored. In this paper, I delineate the sacred dimensions of Ruth as king and savior of baseball. I demonstrate how that narrative fits in the larger American civil religion as it was being transformed from pastoral to urban and expanded from Puritan and Protestant to include ethnic immigrant Catholic and Jewish. Finally, I argue that Ruth's Catholic identity formed the basis for his canonical roles as Patron Saint of Children, repentant sinner, and champion of the poor and weak and how, because of Ruth, those elements of Catholic religion became part of American civil religion.

2. Babe Ruth and the Religion of Baseball

> If sport has become the national religion, Babe Ruth is the patron saint. He stands at the heart of the game he played, the promise of a warm summer night, a bag of peanuts, and a beer. And just maybe, the longest ball ever hit out of the park.
>
> (Montville 2006, p. 367)

There is a robust literature that argues both for and against applying the category of religion to secular phenomena such as sports (Novak 1976; Bain-Selbo and Sapp 2016; Grano 2017; Higgs and Braswell 2004; Prebish 1993). If, however, we agree with current scholarship that asserts that the religion–secular divide is not a useful way to understand complex societies, then deploying the category of religion to explain the power sports have on the human psyche can be a useful tool. We need not argue whether sports are religions to assert that they can achieve the same ends as systems traditionally defined as religions. Sports do what religions do: they help people make meaning in their lives by developing communal allegiances, honoring holy figures, providing a locus for emotions such as ecstasy and despair, and experiencing the sacred in particular objects, stories, and values (Laderman 2009; Chidester 2005; Trothen 2015). Scholars have long recognized that baseball, in particular, fulfills these functions (Novak 1976; Price 2006; Evans and Herzog 2002). And Babe Ruth, as Montville suggests, is baseball's "patron saint."

Ruth gained this status by his prodigious feats of athleticism. Although he began his professional baseball career as a pitcher (and was masterful at that position as well as every other position he played in his youth), his outstanding accomplishment was hitting a baseball for greater distance and with greater regularity than anyone had ever done (Smelser 1993; Leavy 2018). Before Ruth, a home run was an accidental occurrence. Baseball was a game that highlighted the strategies of moving runners from base to base. Scores were low. Ruth changed the game by trying to hit home runs and succeeding beyond the wildest measure (Leavy 2018). Stout (2016) called Ruth's home runs a "wonderful surprise" that sold newspapers and amazed fans. He could do this because he had superior reflexes and eyesight. Sports medicine experts deemed him a "neuromuscular genius" (Leavy 2018, p. 305). He also worked at perfecting his signature craft, and more than likely experienced the ecstasy of athletes scholars describe as "flow" (Csikszentmihalyi 1990) and "pure joy" (Pipkin 2008) as an integral part of this experience that pleased him and those who watched in awe (Leavy 2018).

Ruth's home runs became a spiritual experience that inspired wonder in those who observed him, initially because of the novelty and ultimately because of the majesty. Seeing Ruth hit home runs was also a common occurrence, even before the days of television, because Ruth played baseball not only during the regular season but also before and after, almost all year round. He "barnstormed;" that is, he toured with a team he organized, often called "The Bustin' Babes" and played games against other major leaguers (often Yankee teammate Lou Gehrig's "Larrupin' Lous") and semi-pro and amateur teams (white and black) after the season ended. He also played exhibition games before the season began up and down the east coast, so many admirers in small towns across the country were able to witness him hitting in person. (Many who could not possibly have done so claimed to nevertheless.) Spectators came not to see the game, but to experience a Babe Ruth home run. He rarely disappointed them (Creamer 2005; Montville 2006).

Numbers are sacred in religious traditions. Ruth's home run numbers were no exception. In 1927 he reached his highest season total, 60. That number became sacrosanct, to the extent that when another Yankee, Roger Maris, hit 61 home runs in 1961, there was such an uproar that legend had an asterisk placed next to his accomplishment in "the record book" so as not to detract from Ruth's record even as Maris surpassed it. Maris also received death threats for challenging Babe's immortal record, but nowhere near as many as when a black athlete, Henry Aaron, surpassed Ruth's lifetime record of 714 home runs. The holiness that was attached to Ruth was palpable and powerful. Even his uniform number, 3, became part of the lore, even though it only marked his place in the Yankee batting order and had no particular meaning to him.

His home runs also created a world of sacred objects and spaces. The bats and the balls he willingly autographed in quantity and for whomever asked became not only prized possessions, but relics (Montville 2006). Stories abound of Ruth's generosity in passing these objects on to the boys and men who waited patiently for them and preserved them with the sanctity they deserved (Smelser 1993).

Yankee Stadium was the sacred space that would come to be known as "The House that Ruth Built." It was designed to highlight his talents:

> The Stadium was a grand monument to the drawing powers of the resident right fielder. Did the Romans ever build a stadium simply to show off the talents of one gladiator? And if they did, did they—as the Yankees did—situate the playing surface so the late-afternoon sun always would be behind their star attraction, not shining in his eyes? (Montville 2006, p. 174)

It was also where his body lay in state after his death. Ken Burns' film *Baseball* documents that many New York children believed that Ruth was buried under the stadium monument that was subsequently erected in his memory (Burns et al. 2004). His birthplace and actual grave are also holy pilgrimage sites, as is his birthplace, now a museum in Baltimore, and the Baseball Hall of Fame where visitors still go to see his locker, gloves, bats, balls, and uniforms (Leavy 2018). As with any saint, things he touched were imbued with a sacred power.

Ruth was not only a saint, he was called the king, another expression of his sacred status that marked the connection between royalty and divinity that was more prevalent in that era than it is today. He was frequently depicted in cartoon sketches wearing a crown and ermine cape. He was often referred to as a "demigod." The press came up with alliterative epithets for Ruth that alluded to his power and his regal presence, not unlike the traditions in Judaism and Islam of making elaborate lists of names that express the awesome power and majesty of God. The most famous title for Ruth was "Sultan of Swat" but there were countless more. Here's a sample:

> Colossus of Clouters, Sultan of Swat, Son of Sock, Caliph of Crash, Goliath of Swat, Mastodonic Mauler, Knight of Swat, Master Mauler, Bazoo of Bang, Maryland Mauler, Rajah of Rap, Baron of Bam, Tarzan of the Diamond, and Batterin' Bambino. (Wehrle 2018, p. 55)

Ruth was also widely understood as baseball's savior. He not only transformed the game as played on the field by redefining the role of the home run but also changed the narrative about baseball's role in society. The association between baseball and gambling was prevalent in the early years of the twentieth century. It was not only a rough (and sometimes violent) game on the field, but a corrupt one off the field. Ruth's miraculous home run output in 1920 and 1921 became the story that newspapers reported most frequently, overshadowing accounts of the "Black Sox scandal" in 1919 (Wehrle 2018). Ruth's accomplishments sold more newspapers than the story of several Chicago White Sox players accepting money from gamblers to throw the World Series. In providing a distraction from this story, Ruth changed baseball's reputation and thereby received credit for its salvation (Smelser 1993). In its secular meaning, Ruth definitely saved baseball from the harm this scandal caused. But the Christian meaning of salvation, being saved from sin, also applies here, as gambling on sports was indeed considered sinful, and it was Ruth who provided redemption.

Of course, the true redemption of baseball from the sin of gambling was in the hands of the newly appointed authoritarian commissioner, Judge Kennesaw Mountain Landis, who was hired by team owners to bring a new era and aura to baseball. Part of Landis' task was to end gambling, punish players who broke rules (including Babe Ruth) and make the game more palatable to a middle-class audience. But Ruth's savior role should not be underestimated (Smelser 1993). He changed the game not only with his ability but also with his personality and larger than life self-presentation.

Landis represented authority; Ruth represented joy. As one scholar suggested, "Landis ruled the state; Ruth ruled the people" (Stoloff 2008, p. 91). No doubt, he was baseball's King and the Savior in the eyes of the public.

Ruth's god-like powers on the baseball field also extended to the legendary "curse of the Bambino" that changed the baseball fortunes for the city of Boston and the Red Sox. The curse was punishment for the sin committed by the Boston Red Sox in 1919 when they sold Babe Ruth's contract to the New York Yankees. Popular lore imagines that for 86 years the Red Sox failed to win a World Series because they foolishly cast out the Babe (Shaughnessy 1990). The story has been explained in biblical terms as baseball's original sin where Boston is Eden, Ruth Adam and the Boston owner, Harry Frazee, the serpent (Ardolino 2004). Given that Ruth, like Adam, had no power to determine where he played, and had settled happily in Boston, he was not the villain. Frazee, "the snake," who made the deal, was blamed for endangering the future of the team by selling Ruth for purely financial reasons. That Ruth was difficult to manage may have been part of the consideration as well, but the story is rarely framed to put the blame on him (Stout 2016). Ruth was coveted, not hated by Boston's fans. People believed he had the power to end the curse, even after his death. Rituals at his former home in Sudbury to exhume a piano he legendarily heaved into a lake and countless tales of propitiations and incantations at his gravesite serve to reinforce the Babe's supernatural powers to save Boston as he saved baseball (Shaughnessy 1990).

3. American Civil Religion and Catholic Identity

He will be the patron saint of American possibility. (Montville 2006, p. 13)

Ruth was central to the fabric of American mythology beyond the confines of the baseball diamond. While his abilities made him a baseball god, his personality and life experiences made him, if not a god, a dominant figure in the fabric of the American society of the 1920s. Recovering from the difficulties of the World War and on the cusp of Depression, America needed a singular hero to celebrate. Ruth fulfilled that role. He was a man of extraordinary powers in a country that was beginning to see itself as a world power. By the end of the 1920s, Ruth was rated in polls as the most famous American behind only Washington and Lincoln (Smelser 1993). During the Depression, he joked that he made more money than the President did because he was better at his job (Smelser 1993). He was America's "everyman;" easy to relate to across geographical and class differences. His family was merchant class, and he was raised as working poor, but as baseball's most heralded and best player (and through wise management of his salary) he became rich (Montville 2006). His rise from humble origins to wealthy sports hero made Ruth the paradigmatic Horatio Alger for his time. In religious terms, he exemplified the Protestant myth of the Prosperity Gospel that equated economic success with morality and goodness. Additionally, the style of play that Ruth introduced to baseball demonstrated a Muscular Christianity that equated athletic prowess with Protestant valor.

Ruth, however, did not subscribe to the Protestant civil religious teachings of the prosperity gospel or the muscular Christianity that extolled athletic prowess, hard work, and economic success as signs of piety. Ruth's story is not a celebration of the Protestant values of the American civil religion as he was not Protestant but Catholic. Ruth's Catholic identity brings a new perspective that expands American civil religion to include the religious practices and beliefs of ethnic immigrants.

Although ethnic Catholics had achieved acceptance by the end of the nineteenth century, the Catholic religion was still approached with skepticism and fear throughout the 1920s. Concerns about foreign (papal) influence made people apprehensive about Catholic loyalties to foreign entities. Unfamiliarity with the intricacies of Catholic ritual (often seen as superstitious) and distrust of clerical celibacy also played a role (Smith 2010; Steinfels 2004). Ruth's popularity, however, was part of a process that would make Catholic teachings and rituals an acceptable part of the story of America and provided a new dimension to American civil religion.

Ruth had a strong and very public Catholic identity. Ruth's father was of Lutheran background, but his mother was a practicing Catholic. His parents were occupied with other matters: his father owned a bar and his mother suffered from ill health, and young George was too much for them to handle. He was sent away from home at age seven, and his parents did not have much influence on his upbringing after that (Smelser 1993). Ruth's religious identity was formed as a boarding student at St. Mary's Industrial School run by the Catholic order of St. Francis Xavier. The lay Catholic order of Xaverian Brothers who lived and taught at St. Mary's were his models and mentors. Ruth lived there for twelve years and the Brothers were his legal guardians for his entire youth. He received training in Catholic doctrine and attended Mass on a daily basis. At his confirmation, he took the middle name Herman after Brother Herman, the head of Athletics (Smelser 1993).

Baseball was played widely in Catholic schools at the time because of the influence of Irish Catholic professional ballplayers (Campmier 2018). Baseball played a role in the moral and social development of Catholic youth, teaching them both discipline and skills. Ruth played baseball daily at St. Mary's, and the Brothers were attentive to his superior talents. Brother Matthias, who was in charge of discipline and worked in the athletic program became his mentor (Montville 2006). For the rest of his life, Ruth was connected to the school. He visited regularly, brought their band to play at ballparks where he barnstormed, and supported the school financially (Smelser 1993).

After leaving St. Mary's, Ruth remained devoted not only to the institution, but to his Catholic faith. In Catholicism, he found a set of allegiances to maintain, rules to follow (or not), and rituals to observe that were comfortable and comforting. He was a member of the Knights of Columbus. He became a very public supporter of Catholic philanthropic causes such as Boys Town and the Monterey-Fresno-Catholic Diocesan Campaign (Leavy 2018). He attended Mass and tithed conspicuously, albeit irregularly (Montville 2006). Both of his weddings took place in the Church. He stayed married to his first wife Helen, even though they were estranged, obeying Catholic rules against divorce. When Helen died tragically, Ruth demonstrated his grief immediately by attending mass at St. Cecilia's Church where he was seen "fingering his big brown rosary which has numerous relics from the Shrine of St. Ann de Beaupre enclosed on cross" (Wehrle 2018, p. 187). Whenever Ruth found himself in trouble for insubordination on the baseball field or for his pursuits of pleasure off the field, a Catholic figure of note, often with ties to St. Mary's, was brought in to counsel and support him (Montville 2006).

It is also possible that Ruth's one foray into politics, his open support of Al Smith's Presidential campaign in 1928, had something to do with his affinity for Smith as a Catholic whom he saw as a kindred spirit. He was publicly critical of those who opposed Smith on the grounds of "petty prejudice" and admired his stance against Prohibition (Wehrle 2018).

By the mid-1930s, Catholicism had become more widely accepted in American popular culture (Smith 2010). At the same time, Ruth's baseball career was in decline. When his prodigious abilities began to wane with age, the Yankees no longer needed him. They refused to grant his wishes to manage the team, claiming he did not have the skills required to lead (Wehrle 2018). He ended his career with the Red Sox and then later worked with the Brooklyn Dodgers as an assistant coach, retiring in 1935. But he could no longer hit and was no longer the box office draw (Leavy 2018) and the experiences ended badly. He spent his last years doing charitable work, lending his name to and traveling on behalf of a variety of causes. Cancer cut short his life after baseball. As he was dying, baseball embraced him once again with tributes and accolades recalling his glorious career (Leavy 2018). In these final years, Ruth claimed to have relied on Catholicism to "get [his] house in order" (Ruth and Considine 1948, p. 230).

Ruth's death, funeral, and burial were conducted with Catholic rites and traditions. He kept a statue of Martin de Porres (patron saint of mixed-race people and founder of orphanages) at his hospital bedside during his last hospital stay, and many rosaries and pennies for mass arrived from fans who wished him well. He received last rites

> from Father Thomas H. Kaufman, which some members of the faith felt he didn't deserve. Kaufman, a Dominican priest from St. Catherine of Siena parish, was filling in for a priest who was on vacation. He was from Baltimore and had spent one troubled night at St. Mary's as a boy. (Leavy 2018, p. 458)

After his death on 16 August 1948 Ruth lay in state at Yankee Stadium, "with black rosary beads wound around his thick fingers" and a massive crucifix and vigil candle at one end of the coffin (Leavy 2018, p. 468). There were so many fans (estimates ranged from 77,000–250,000) who wanted to file past his coffin to say good-bye that the Stadium was kept open overnight and into the next day to accommodate them (Leavy 2018).

Ruth's funeral took place at St. Patrick's Cathedral on 20 August, with Cardinal Francis Spellman and a retinue of forty-four priests presiding. Newspapers covering the event described the solemn ritual in detail, including how "the Cardinal descended from his throne, walked slowly to the foot of the coffin, and gave the final absolution" (Feinberg 1948).

Newspapers reported that over 100,000 people lined the funeral route to Ruth's final resting place at the Catholic Cemetery, Gate of Heaven, in Valhalla, New York. Jane Leavy described his tombstone:

> There's nothing subtle or understated about the gravesite, dwarfed as it is by a sandblasted statue in Westerly granite of Jesus blessing a young baseball player. It is engraved with the words of Cardinal Spellman: 'May the Divine Spirit That Animated Babe Ruth to Win the Crucial Game of Life Inspire the Youth'. (Leavy 2018, p. 476)

Ruth's grave tells the story of his life as an American Catholic. His spirit lives on in Spellman's words and is contained in both the images of the boy and the Christ. By the time of Ruth's death, the pomp of Catholic religious ceremony had been integrated into the fabric of American life and Ruth may be credited in part for the level of acceptance it achieved.

4. Ruth's Catholic Commitments

Babe Ruth was a public celebrity, a hero for the ages. Some of his fame came from his prodigious accomplishments in a game that he helped to elevate the national pastime and American society's growing interest in sports. Some of his fame was because this transpired as America transitioned from a rural to an urban centered society while Ruth was living and working in the greatest urban area, New York. His broad popularity can be attributed to the new media that made Ruth accessible: talking pictures, newsreels, newspaper images, and the sports pages that featured daily articles about the Babe's exploits by some of the leading writers of the day, including Damon Runyan, Bob Considine, and Paul Gallico, not to mention Ruth's own ghost-written columns. Scholars, most notably Leavy (2018), have focused on the contribution of his agent, Christy Walsh, who was responsible for turning Ruth into a celebrity in the new mediated consumer culture. Part of Walsh's genius was presenting Ruth in familiar mythic poses that emphasized Ruth's status as baseball's king and savior.

While the religious motifs made Ruth a god of the religion of baseball, his public identification as a Catholic brought a new dimension to the civil religion of America. Ruth's main religious significance lies in three key aspects of his life, expressed in the mythic language of Catholic traditions. As his tombstone suggests, Ruth functioned as the patron saint of children, best illustrated through the Johnny Sylvester legend. His propensity towards immature behavior, pleasure seeking, and public apology corresponds to the Catholic pattern of sin and repentance. The ways Ruth challenged authority when he thought he or others were being treated unfairly made him a champion of economic justice and a proponent of what came to be known as the preferential option for the poor, a newly developing dimension of Catholic doctrine.

5. Patron Saint of Children

George Herman Ruth's nickname was bestowed on him when, as a teen, he left St. Mary's to play professional baseball. His new guardian, Jack Dunn, was described as having this "babe" playing for him (Smelser 1993). The name stuck and perpetuated an overemphasis on Ruth's childlike demeanor (Wehrle 2018). Close associates called him "Jidge" or sometimes less flattering names (Montville 2006). He called everyone "kid," perhaps to level the playing field (Creamer 2005). But as much as Babe was the baseball playing child represented on his tombstone, he was also the Christ-like figure who cared for and blessed children and was credited with healing powers.

Ruth took his role as patron saint of children very seriously. In his ghost-written columns, he sometimes took on the role of moral counsellor, offering advice to America's youth that often focused on fair play and teamwork (Wehrle 2018). He cheerfully autographed baseballs with the beautiful penmanship he learned at St. Mary's (Montville 2006). He seemed happiest surrounded by boys who saw him as their hero. His visits to children in orphanages and hospitals were well documented and achieved the status of legend (Montville 2006).

There are many stories of children being restored to health through Ruth's magic touch (Lloyd 1976). Ruth's connection to children transitioned to sainthood when he performed a "miracle" for Johnny Sylvester (Leavy 2018). The story brought heightened attention to Ruth's large and magnanimous relationship to poor and ailing children by encapsulating it in a story that highlighted its mythic dimensions even though it featured a well-to-do and not very ill boy. As with all myths, this one contains a kernel of truth. But its embellishment provides a more interesting window into the religious significance of Babe Ruth for American culture and illustrates how Ruth helped to integrate and normalize Catholic religious ideas such as miracles, once assumed to be mere superstitions, into America's civil religion.

Johnny Sylvester was an eleven-year-old boy from a wealthy family who had been thrown from a horse and suffered complications while recovering at home in Essex Falls, New York. Johnny was a big baseball fan. Knowing that autographed baseballs would cheer his son up, his father contacted friends in St. Louis where the Yankees were about to play the Cardinals in the 1926 World Series. Ruth's publicists sent balls signed by the Yankees and Cardinals with a note from Ruth promising Johnny he would hit a home run for him. When Ruth hit three World Series home runs, the story of Ruth's keeping a promise to a sick boy began to circulate in the press. The story escalated when the reports began to allege that Johnny made a miraculous recovery as a result of Ruth's intervention. After the World Series, *The Daily News* kept up the attention, by arranging for Ruth to go visit Johnny as he recovered. Ruth happily complied, stopping in Essex Falls on his way to a barnstorming game against the Negro League Brooklyn Royal Giants (Wehrle 2018; Leavy 2018).

Johnny Sylvester's recovery went down in baseball annals as a miracle attributed to Ruth. The story was fashioned in ways similar to miracles that are the requirement for sainthood in the Catholic Church. For a miracle to be attributed to a Catholic who is nominated for beatification, a team of experts must certify that miraculous recovery was based on scientific evidence and attributable to prayers made to the would-be saint. Sylvester himself achieved fame as "Babe Ruth's kid" as his obituary attests (Thomas 1990). He was brought to Ruth's bedside when the Babe was dying to further embellish the legend and make the promise of healing reciprocal.

The story of Babe Ruth's home runs and their saving power became an even more powerful myth when Hollywood took it over. It found its way into *Pride of the Yankees* (Goldwyn et al. 2008), the film about Ruth's rival and teammate, Lou Gehrig. *Pride of the Yankees* was made in 1942 immediately after Gehrig's death from what came to be known inaccurately as "Lou Gehrig Disease." The film was nominated for the Academy Award for best picture, and starred Gary Cooper as Gehrig while Ruth played himself quite credibly. In 1942, Ruth was a household word, and Gehrig not as well known. Having Babe in the film sold tickets. And, not surprisingly, the film adapted the Johnny Sylvester miracle story to suit its purposes. The scene where Ruth and Gehrig visit "Billy" in the hospital before a World Series game makes Gehrig, not Ruth, the hero. Ruth promises to hit a home run for Billy, but Gehrig additionally engages with the boy, encouraging him, and promises two home runs that, of course, he hits. When Gehrig is dying, Billy shows up to inspire him and let him know that thanks to Gehrig's encouragement he learned to walk again.

Hollywood embellished and degraded the legend even more in the 1948 *Babe Ruth Story* (Del Ruth 1948) that was made just before Ruth died. Here, the Sylvester miracle is conflated with Ruth's other "miracle" home run, known as "the called shot." Legend has it that Ruth pointed to the outfield (center or right) and then hit a home run to that spot in the 1932 World Series against Chicago. No one can actually recall seeing this occur (including President Roosevelt who attended the game) and Ruth would never affirm or deny having predicted the shot (Ruth and Considine 1948). In the film, the Johnny character is in the hospital, dying, hears on the radio that Ruth pointed and hit that home run for him, and recovers. The film attributes other miracles to Ruth and his healing powers with children that do not even have the kernel of truth necessary for a myth to take root. The film is considered by critics among the worst movies ever made (Montville 2006) and reduces Babe to a well-meaning boob, not the saintly patron of children that he actually was. Nonetheless, Ruth's miracles for children made this Catholic belief a more acceptable part of American civil religion.

6. Repentant Sinner

Was Ruth, as Smelser (1993) aptly describes the sides of Ruth's personality, more the "Baltimore waterfront slob" or the "Xaverian Brother George"? Recent scholarship has shown that tales of his hedonistic behavior were wildly exaggerated (Wehrle 2018). Ruth began his life as the "Baltimore waterfront slob" and as Ruth's world broadened from his Baltimore roots, he had to learn to deal with fame and fortune. The Catholic paradigm of repentant sinner describes the process Ruth frequently underwent as he learned and experimented, and then saw (or was made to see) the error of his ways.

Leigh Montville described Ruth's pattern this way:

> The Catholic religion would stay with him, the rhythm of mistakes and redemption perfect for his life of rapidly accumulated venial sins. Three Our Fathers, three Hail Mary's, and a good Act of Contrition would clear out his moral digestive system and set him back on the road. He would amaze teammates sometimes when he would appear at Mass in the Morning after a night of indulgence. Three Our Fathers, three Hail Marys, a good Act of Contrition, a $50 bill in the collection basket, ready to go. (p. 29)

The pattern of Ruth attaining a god-like status in the eyes of his fans (worshippers) and then failing to live up to expectations followed him throughout his career. While still with the Red Sox, he got into a fight with the manager, Ed Barrow, about his late-night carousing. After Barrow suspended him, "a chagrined Ruth went to his manager as if he was going to confession at St. Mary's and apologized" (Stout 2016, p. 120). In 1922, he fought publicly with Judge Landis over breaking a rule about barnstorming, and ended up apologizing and accepting his punishment. The rest of the season was also difficult for Ruth, who was suspended and fined several more times for arguing with umpires and other displays of temper. He played poorly in the World Series, and the popular politician, soon-to-be mayor of New York, Jimmy Walker, scolded him publicly for letting down the youth of America. His response to Walker was to weep openly and promise to set a better example (Smelser 1993).

The most dramatic episode in this cycle of sin and repentance took place in 1925. Ruth came to spring training seriously out of shape and overweight. He suffered from a malady that sports writers variously identified as a groin injury, an infection, syphilis, and alcohol-related behavioral problems. They called it "the bellyache heard round the world" and attributed it to his overindulgence. Whatever the cause, he collapsed during spring training and was taken by train to New York where he was hospitalized at St. Vincent's for six weeks. He did not play regularly until June and did not play up to the levels he had achieved earlier in his career. At age 30, it was rumored that his good years were behind him. Yet when he returned, he played well if not up to his earlier prodigious levels. But he also fought with his manager, Miller Huggins, refused to follow curfew rules, and often arrived late and sullen to games. Late in the season, Huggins suspended him, and Ruth did not react well. In response to his protests, the Yankee front office supported Huggins, not their star player. Again, Ruth the sinner took the role of penitent quite seriously. He and his advisers brought Brother Matthias to counsel him. Ruth swore to reform his behavior, and publicly presented his mentor with a Cadillac to thank him for raising him to be "Xaverian Brother George" (Leavy 2018).

Ruth truly took his repentance to heart. He spent the winter in a new scientifically-based health regimen at the McGovern Gym in New York City that got him into shape for the 1926 season. He arrived at spring training in excellent condition and began to play as or more magnificently than he had in his peak years from 1921–1924. His agent, Christy Walsh, crafted many of Ruth's ghost-written columns as apologies for his misbehavior, describing at length how Ruth was taking responsibility for his actions. With his redemption came an even greater commitment to boys in poverty (Leavy 2018). Among other efforts, he got involved in the Father Flanagan campaign to keep farmers from hiring poor boys in harvest seasons and then not compensating them (Wehrle 2018).

Ruth was a fallen hero, but playing on the theme of repentance and with the support of good advisors such as Walsh and Claire Hodgson (who would later become his second wife) he was able to redeem himself. Although many writers (and the Yankee organization) still viewed him as "the Baltimore waterfront slob" and bad boy, he was, in the eyes of the many who saw him as the god of baseball, truly an example of a good, repentant Catholic, "Xaverian Brother George." Ruth introduced this cycle of sin and repentance that one might call "muscular Catholicism" into the vocabulary of American civil religion.

7. Babe Ruth and the Catholic Doctrine of Preferential Option for the Poor

In the second half of the twentieth century, Catholic liberation theology developed the doctrine of a preferential option for the poor. Following the teachings of the Prophets of the Hebrew Bible and Jesus in the Gospels, Catholic ethicists argued that because God favored the poor and the weak (often represented by widows and orphans), humans were obligated to do so as well. Babe Ruth's actions made him a forerunner of these Catholic ideals. Although in American civil religion, Ruth represented the Protestant gospel of prosperity, Ruth's behavior towards blacks, women and the poor fit more clearly with this Catholic doctrine. This emphasis on doing for others was the antithesis of the gospel of prosperity that encouraged economic success for one's own salvation. Ruth also questioned the authority of baseball's overlords. To Jane Leavy, he was the "rule breaker in chief" (p. 102).

There is no doubt that Ruth was widely loved by Americans as the king and savior of baseball. But he had his detractors, especially those among baseball's powerful who saw Ruth as the sinner, the "Baltimore waterfront slob," who had to be kept in line. The journalists and players who criticized him mocked not only his class background, but also questioned the purity of his whiteness. Ruth never backed away from his roots in poverty. Although there is no evidence to suggest that Ruth came from African stock, he often expressed his affinity with America's blacks (Leavy 2018).

Ruth was frequently the subject of racial invective. Along with the nicknames that emphasized his divine powers, there were other epithets hurled at him that suggested that Ruth came by those powers because he was of African descent, and, according to racial stereotypes, had beast-like power. He was called N—Lips by teammates in Boston in reference to his wide nose, thick lips, and swarthy complexion. The innuendos around his orphan status (that he always denied), his inability to control his emotions, and his sexual prowess supported racist stereotypes (Montville 2006). One minister in Boston connected the curse of the Bambino with New England racism, arguing that Boston should have paid for that sin (Shaughnessy 1990).

Ruth did not like the taunts, but he would not be cowed by them. Nor did it affect his approach towards African Americans with whom he maintained close relationships throughout his life. While he was far from a civil rights activist, he publicly supported black causes and institutions. He lent his name to fundraising efforts for black churches and frequently visited segregated orphanages and hospitals (Wehrle 2018). More than other white players of the era, he barnstormed against (and occasionally played with) Negro League teams, befriended their players, and spoke highly and publicly of their baseball abilities (Leavy 2018).

Ruth also demonstrated respect for women. While he exemplified muscular Christianity to America, his own masculinity was not of the muscular kind. Although raised in the exclusive company of men and boys as St. Mary's, Ruth clearly preferred the company of women as an adult. While he was not faithful to his first wife, most of his sexual exploits remained unsubstantiated (Wehrle 2018). He respectfully refused to divorce Helen for reasons of both religion and compassion. Like Ruth's mother, Helen was not mentally stable. When they separated, he continued to provide for her and the daughter she (likely) adopted when they were estranged. He had the utmost respect for his second wife, Claire Hodgson, and her mother and daughter with whom Ruth remained close (Montville 2006).

Claire helped Ruth "repent." Under her guidance, he became a fully developed model of "Xaverian Brother George." Once married, she set him on a regimen of healthy living that he willingly obeyed. The press sometimes depicted him as a boy who had found a mother to care for him (Wehrle 2018). As he settled down during his second marriage, they could no longer deploy the male-dominant narrative of Ruth as profligate adventurer, drunk, and playboy. In this marriage, Ruth displayed a willingness to challenge gender norms. It is possible that this willingness to cede the upper hand to his wife was a factor in Ruth not achieving the goal he had set for himself, to become a major league manager, as owners could portray him as childishly subordinating himself to a woman (Wehrle 2018).

When a woman pitcher, Jackie Mitchell of the Chattanooga Lookouts, struck Ruth out during an exhibition game, Ruth was a willing participant in the story. As the *New York Times* reported, he swung and missed at two balls, asked for the ball to be inspected, "just as batters do when utterly baffled by a pitcher's delivery." He struck out on a called third strike, "flung his bat away and trudged to the bench, registering disgust with his shoulders and chin" (Brandt 1931, p. 32). Whether or not the event was a publicity stunt, Ruth played his role. Allowing a woman to best him in public required a willingness to challenge the stereotypes of gender and a chivalrous acceptance of a different kind of masculinity.

Finally, Ruth's preference for the poor manifested itself in his continuous struggles with baseball's owning class over the rights of players to control their own economic destinies. He was an early supporter of a players' union when he first entered professional baseball. He fought annually with management in Boston and New York over his compensation (Smelser 1993). Although Ruth got a handsome salary, it was clear that team owners profited even more handsomely from Ruth. He was primarily responsible for building the popularity of the game and that translated into gate and residual receipts vastly higher than what he was paid (Wehrle 2018). He made more money off-season selling his name than he did during the season as a player (Leavy 2018). He rebelled against a reserve system that treated players like indentured servants, tying them to teams for life contracts if the team so chose. He also criticized his fellow players when they did not treat each other well, as in 1932 when Chicago Cubs refused to share their bonuses equitably. As Edmund Wehrle concludes:

> Compared to athletes in other sports, Ruth was underpaid. No one could reasonably argue that Ruth received a fair share of the income he generated for his sport. Ruppert, as team owner, and Barrow, as team secretary and owner of a share of the Yankees, profited immeasurably from Ruth. With some cause, Ruth carried a working-class view of the world and a strong impulse to defend the value of his labor and the value of baseball players in general. (p. 166)

8. Conclusions

Looking at George Herman "Babe" Ruth from the perspective of religion, we find a baseball god and exemplar of Catholic values. Ruth recreated the game of baseball with his majestic powers, inspiring awe and love from fans. The home runs he hit forever changed the way the game was played. Ruth died over half a century ago, yet his name remains a household word and the symbol of baseball's importance in society. As the paradigmatic American hero of his time, he also changed the view of Catholic rites and rituals, helping to normalize Catholic religious beliefs and making them an acceptable part of American civil religion. He brought attention to Catholic values of love for children, forgiveness, and support for economic and social justice that are still important today and should forever be associated with the King of Baseball, Babe Ruth.

Funding: This research received no external funding.

Conflicts of Interest: The authors declare no conflict of interest.

References

Ardolino, Frank R. 2004. The Curse of the Bambino (review). *NINE: A Journal of Baseball History and Culture* 13: 114–17. [CrossRef]

Bain-Selbo, Eric, and D. Gregory Sapp. 2016. *Understanding Sport as a Religious Phenomenon: An Introduction*. New York: Bloomsbury.

Brandt, William E. 1931. Special to *The New York Times*. Girl Pitcher Fans Ruth and Gehrig. *New York Times*. April 3. Available online: http://libproxy.temple.edu/login?url=https://search.proquest.com/docview/99362273?accountid=14270 (accessed on 13 May 2019).

Burns, Ken, Lynn Novick, and Geoffrey C. Ward. 2004. *Baseball*. Alexandria: PBS Home Video, Disc 4.

Campmier, David M. 2018. Professional Baseball's Emerald Era: Irish Catholics and Early Major League Baseball, 1880–1910. *U.S. Catholic Historian* 36: 55–73. [CrossRef]

Chidester, David. 2005. *Authentic Fakes: Religion and American Popular Culture*. Berkeley: University of California Press.

Creamer, Robert W. 2005. *Babe: The Legend Comes to Life*. New York: Simon & Schuster.

Csikszentmihalyi, Mihaly. 1990. *Flow: The Psychology of Optimal Experience*. New York: Harper & Row.

Del Ruth, Roy. 1948. *The Babe Ruth Story*. Burbank: Warner Home Video.

Evans, Christopher Hodge, and William R. Herzog, eds. 2002. *The Faith of Fifty Million: Baseball, Religion, and American Culture*. Louisville: Westminster John Knox Press.

Feinberg, Alexander. 1948. 75,000 Go to Babe Ruth's Funeral and Stand in Rain along Fifth Ave. *New York Times*. August 20. Available online: http://libproxy.temple.edu/login?url=https://search.proquest.com/docview/108395648?accountid=14270 (accessed on 13 May 2019).

Goldwyn, Samuel, Paul Gallico, Jo Swerling, Herman J. Mankiewicz, Sam Wood, Gary Cooper, Teresa Wright, Walter Brennan, Babe Ruth, Bill Dickey, and et al. 2008. *The Pride of the Yankees*. Beverly Hills: Distributed by 20th Century Fox Home Entertainment.

Grano, Daniel A. 2017. *The Eternal Present of Sport: Rethinking Sport and Religion*. Philadelphia: Temple University Press.

Higgs, Robert J., and Michael Braswell. 2004. *An Unholy Alliance: The Sacred and Modern Sports*. Macon: Mercer University Press.

Keane, Robert N., ed. 2008. *Baseball and the "Sultan of Swat": Babe Ruth at 100*. New York: AMS Press.

Laderman, Gary. 2009. Sacred Matters: Celebrity Worship, Sexual Ecstasies, the Living Dead, and Other Signs of Religious Life in the United States. New York: The New Press.

Leavy, Jane. 2018. *The Big Fella: Babe Ruth and the World He Created*. New York: Harper Collins Publishers.

Lloyd, F. R. 1976. The Home Run King. *The Journal of Popular Culture* 9: 983–95. [CrossRef]

Montville, Leigh. 2006. *The Big Bam: The Life and Times of Babe Ruth*. New York: Doubleday.

Novak, Michael. 1976. The Joy of Sports: End Zones, Bases, Baskets, Balls, and the Consecration of the American Spirit. New York: Basic Books.

Pipkin, James W. 2008. *Sporting Lives: Metaphor and Myth in American Sports Autobiographies*. Columbia: University of Missouri Press.

Prebish, Charles S. 1993. *Religion and Sport: The Meeting of the Sacred and the Profane*. Westport: Greenwood Press.

Price, Joseph L. 2006. *Rounding the Bases: Baseball and Religion in America*. Macon: Mercer University Press.

Ruth, George Herman, and Bob Considine. 1948. *The Babe Ruth Story*. New York: E. P. Dutton.

Shaughnessy, Dan. 1990. *The Curse of the Bambino*. New York: Dutton.

Smelser, Marshall. 1993. *The Life That Ruth Built: A Biography*. Lincoln: University of Nebraska Press, Available online: http://search.ebscohost.com/login.aspx?direct=true&db=nlebk&AN=44577&site=ehost-live&scope=site (accessed on 13 May 2019).

Smith, Anthony. 2010. *The Look of Catholics: Portrayals in Popular Culture from the Great Depression to the Cold War*. Lawrence: University of Kansas Press.

Steinfels, Margaret O'Brien, ed. 2004. *American Catholics, American Culture: Tradition and Resistance*. Lanham: Rowman & Littlefield.

Stoloff, Sam. 2008. The Sultan and the Czar: Babe Ruth, Judge Landis, and the Carnivalesque in Mass Culture. In *Baseball and the "Sultan of Swat": Babe Ruth at 100*. Edited by Robert N. Keane. New York: AMS Press, pp. 87–94.

Stout, Glenn. 2016. *The Selling of the Babe: The Deal that Changed Baseball and Created a Legend*. New York: Thomas Dunne Books, St. Martins Press.

Thomas, Robert McG., Jr. 1990. Johnny Sylvester, the Inspiration for Babe Ruth Heroics, Is Dead. *New York Times*. January 11. Available online: http://libproxy.temple.edu/login?url=https://search.proquest.com/docview/108623799?accountid=14270 (accessed on 13 May 2019).

Trothen, Tracy J. 2015. *Winning the Race?: Religion, Hope, and Reshaping the Sport Enhancement Debate*. Macon: Mercer University Press.

Wehrle, Edmund F. 2018. *Breaking Babe Ruth: Baseball's Campaign Against Its Biggest Star*. Columbia: University of Missouri Press, Available online: http://ebookcentral.proquest.com/lib/templeuniv-ebooks/detail.action?docID=5379870 (accessed on 13 May 2019).

MDPI

St. Alban-Anlage 66

4052 Basel

Switzerland

Tel. +41 61 683 77 34

Fax +41 61 302 89 18

www.mdpi.com

Religions Editorial Office

E-mail: religions@mdpi.com

www.mdpi.com/journal/religions